Teubner-Reihe UMWELT

B. Hock/D. Barceló/K. Cammann/
P.-D. Hansen/A. P. F. Turner (Eds.)

Biosensors
for Environmental Diagnostics

Teubner-Reihe UMWELT

Herausgegeben von
Prof. Dr. mult. Dr. h.c. Müfit Bahadir, Braunschweig
Prof. Dr. Hans-Jürgen Collins, Braunschweig
Prof. Dr. Bertold Hock, Freising

Diese Buchreihe ist ein Forum für Veröffentlichungen zum gesamten Themenbereich Umwelt. Es erscheinen einführende Lehrbücher, Monographien und Forschungsberichte, die den aktuellen Stand der Wissenschaft wiedergeben.

Das inhaltliche Spektrum reicht von den naturwissenschaftlich-technischen Grundlagen über umwelttechnische Fragestellungen bis hin zu juristisch, sozial- und gesellschaftswissenschaftlich ausgerichteten Titeln. Besonderer Wert wird dabei auf eine allgemeinverständliche, dennoch exakte und präzise Darstellung gelegt. Jeder Band ist in sich abgeschlossen.

Die Autoren der Reihe wenden sich vorwiegend an Studierende, Lehrende sowie in der Praxis tätige Fachleute.

Biosensors
for Environmental Diagnostics

Edited by

Prof. Dr. Bertold Hock
Technische Universität München

Prof. Dr. Damià Barceló
Department of Environmental Chemistry Barcelona

Prof. Dr. Karl Cammann
Institut für Chemo- und Biosensorik Münster

Prof. Dr. Peter-D. Hansen
Technische Universität Berlin

Prof. Dr. Anthony P. F. Turner
Cranfield University

 B. G. Teubner Stuttgart · Leipzig 1998

Prof. Dr. Bertold Hock
Fax: +49 81 61 71 44 03
E-mail: hock@weihenstephan.de

Prof. Dr. Damià Barceló
Fax: +34 32 04 59 04
E-mail: dbcqam@cid.csic.es

Prof. Dr. Karl Cammann
Fax: +49 25 19 80 28 02
E-mail: cammann@uni-muenster.de

Prof. Dr. Peter-Diedrich Hansen
Fax: +4 93 08 31 81 13
E-mail: pd.hansen@tu-berlin.de

Prof. Dr. Anthony P. F. Turner
Fax: +44 12 34 75 24 01
E-mail: a.p.turner@cranfield.ac.uk

Gedruckt auf chlorfrei gebleichtem Papier.

Die Deutsche Bibliothek – CIP-Einheitsaufnahme

Biosensors for environmental diagnostics /
ed. by Bertold Hock . . . –
Stuttgart ; Leipzig : Teubner, 1998
  (Teubner-Reihe Umwelt)
  ISBN 978-3-8154-3540-3   ISBN 978-3-322-93454-3 (eBook)
  DOI 10.1007/978-3-322-93454-3

© B. G. Teubner Verlagsgesellschaft Leipzig 1998

Umschlaggestaltung: E. Kretschmer, Leipzig

Preface

Biosensors combine the power of microelectronics with the selectivity and sensitivity of biological components such as whole cells, organelles or biomolecules, e.g. antibodies, receptors, enzymes and nucleic acids. They are used to detect individual substances or groups of substances in the environment, such as industrial emissions that originate, for instance, from the textile, cellulose and pharmceutical industry as well as from agricultural activities. The biosensor approach is expected not only to provide a significant contribution to measurement technology but also a basis for competent political decisions.

Up to now disturbances in lakes and rivers are detected more or less by chance. Essentially, only substances that can be assayed by traditional physico-chemical techniques are found. However, the pollution peak has usually already passed by the time the results become known and acquisition of evidence for the identification of the responsible party is hardly possible after the event. Therefore fast and continuous measurement systems such as biosensors are required to provide inexpensive and cost effective event-related sampling of water thus providing for the preservation of evidence.

Biosensors can detect biological effects such as genotoxicity, immunotoxicity and endocrine responses. The sequence of these signals and especially the peak values provide valuable indicators for water protection and facilitate the elaboration of new strategies and concepts within water management. Of major importance is the knowledge of distribution of warning signals within space and time from water ecosystems. Only then can the "health status" of water be recognized sufficiently early. Biosensors provide a particularly important instrument for control of domestic waste water treatment and ensuring the success of sanitary measures.

This book originated from the 5th European Workshop on Biosensors for Environmental Monitoring and Stability of Biosensors, coorganized by the Technical

6

University of München at Freising-Weihenstephan and the European Commission. However, it is not intended as a proceedings volume; it rather reflects the current knowledge and trends in the field of biosensors for environmental monitoring. We would like to thank the European Commission for its support, the Teubner-Verlag for its kind cooperation and Stefanie Rauchalles and Dr. Karl Kramer for their help editing this book.

January 1998   The Editors

Contents

8

3 Pesticides

4 Monitoring of toxic effects

5 Monitoring of genotoxicity

6 Field experiments

7 Perspectives

1 New Techniques and Analytes

1.1 Resonant Energy Transfer Detection for Low Volume Immunoassay in Environmental Applications

Andreas Brecht, Uwe Schobel, Günter Gauglitz

Institute of Physical Chemistry, University of Tübingen, 72076 Tübingen, FRG

Abstract. Environmental monitoring activities aim at the assessment of the environmental situation and its change due to anthropogenic influence or natural factors. Ideally the environmental picture derived from monitoring activities should be tight in space and time. On the other hand economic factors limit the number of samples which can be processed and analysed. Rapid and cost effective pre-screening techniques are required to reduce the number of samples, which are subject to costly instrumental analysis. Immunoassays are already recognised as a mass screening technique. Immunoassays with reduced sample volume and high sample density may become an answer to the demand for cost effective high throughput screening requirements. Approaches to low volume assays may be adapted from fields like molecular biology, where limited sample amounts or high reagent costs drive similar developments.

This project targets the development of a nanotitreplate immunoassay for environmental analytes. In this paper we focus on the general concept and on results from the immunoanalytical system. The assay volumes foreseen are 100 nl to 500 nl. Microstructured assay compartments (nanotitre plates) are readily available by a variety of techniques (e.g. isotropic or anisotropic etching of silicon, injection moulding, laser manufacturing). Sample and reagent handling will be done by piezo-microdrop techniques, allowing the accurate dispensing of volumes below 1 nl. The small assay volumes foreseen preclude typical sequential ELISA protocols, where washing steps are mandatory. Therefore a wash free proximity type assay is under development. The detection is based on fluorescence Resonant Energy Transfer (RET). A donor-acceptor pair with two "red" (long wavelength excitable) fluorescent dyes is used. This reduces fluorescence background and allows the use of semiconductor laser light sources.

A model assay for atrazine was set up by using an antibody labelled with donor molecules and a conjugate of bovine serum albumin with hapten and acceptor. Initial calibrations gave a test midpoint of 6 ppb.

1.1.1   Introduction

Instrumental techniques are well established in environmental monitoring. Although significant progress has been made, the cost per sample and analyte determined is still considerable. Screening techniques, which allow high numbers of samples to be tested for potential contamination might relax this situation. Immunoassays are recognised as a potentially cheap technology for medium to high sample throughput. Costs per sample and analyte in the clinical field are about 5 ECU. In the environmental field typically a multitude of analytes must be monitored and for screening of samples still lower costs are desirable. One potential way to improve the efficiency of immunoanalytical techniques is to rigorously reduce sample volume, while increasing sample density.

1.1.1.1   Concept

In a joint project ("LINDAU" - see Acknowledgements) funded by the German Ministry of Research, Education, and Technology the potential of a nanotitreplate immunoassay for environmental screening is being investigated. The target is to develop a low volume immunoassay for screening environmental samples. The characteristics of the approach are:

- A microstructured carrier with cavities of 100 nl to 1 µl volume arrayed in a grid with a spacing of 0.5 to 1.5 mm. This allows 50 to 400 cavities per cm².
- A wash-free proximity type detection system using fluorescence labels.
- Contactless sample handling based on piezo microdrop dispensing.
- An imaging detection system, capable of simultaneously or sequentially reading a complete nanotitreplate.

1.1.1.2   Sample carrier

Assay volumes between 100 nl and 1 µl require cavities with dimensions ranging from a few 100 micron to about 1 mm. These dimensions are readily accessible by a broad range of techniques. Anisotropic etching of silicon is one particularly attractive technique in this field. Nanotitreplates machined in silicon are already commercially available with volumes ranging from 500 nl to below 100 nl. Cavities prepared by anisotropic etching assume the characteristic shape of an inverted truncated pyramid. Either by appropriate masking layers or by bonding techniques, cavities with transparent windows can be generated, which allow trans-illumination for fluorescence detection.

1.1.1.3   Microdrop sample handling

Sample volumes between 1µl and 1ml typically are transferred by pipetting. For sample volumes below 1µl contactless transfer techniques are preferable. The use of ink-jet techniques is discussed in this field (Blanchard et al. 1996). Especially the generation of microdrops out of a nozzle of a few ten microns width by pressure waves induced by piezo actuators seems promising. Initially these techniques were developed for printing applications, but the characteristics of the process make it also suitable for handling of samples and reagents in the pl to nl range. Droplets are generated by piezo printing systems (Figure 1) with diameters ranging from 10 µm (< 1 pl) to 100 µm (>1 nl). The droplet volume is reproducible within 1% to 2%. Droplets leave the nozzle with a speeds in the m/s range. This makes the deposition of droplets at predefined positions feasible.

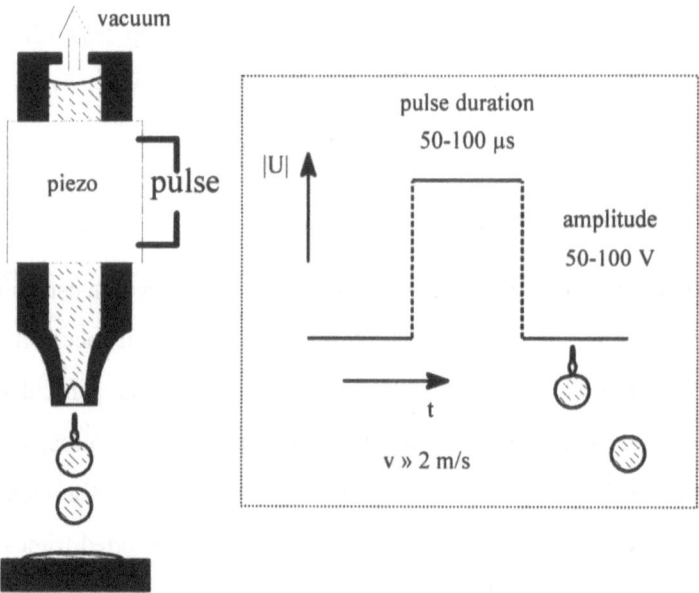

Figure 1: Microdrop dispensing by piezodrop ink-jet technology. Left: inkjet dispenser, right: voltage pulse applied to dispense droplet

1.1.1.4   Proximity type signal generation

Typical immunoassays are based on binding between antibody on the one side and analyte and a competing tracer on the other side. This binding step is followed by a subsequent separation step where "bound" and "free" fraction of the tracer are separated. In ELISA type assays this separation is achieved by binding of reagents to the wall of a microtitre plate, followed by a simple washing step. The use of small cavities in high density formats makes washing steps impossible. Also the repeated addition of reagents in a timed sequence is not desirable. Ideally the assay should be *wash free* and require only the initial mixing of sample and as few reagents as possible. Signal generation should take place without further manipulation. The concept of equilibrium based *proximity assays* is well suited to these needs.

Table 1:   Proximity type affinity assays. A detectable quantity - eg. a fluorescence signal - is directly modulated by the binding of a tracer to the antibody

Assay type	Principle	Literature
Fluorescence quenching	A fluorophore which is sensitive to its local environment is conjugated to an analyte derivate. The fluorescence of the tracer is quenched upon binding to the antibody.	Lin et al. 1997
Fluorescence polarisation (FPIA)	A fluorophore/analyte conjugate is used as tracer. The rotational diffusion of the tracer is decreased upon binding to the antibody. By excitation with a linearly polarised light source, this can be detected as an increase in fluorescence anisotropy.	Dandliker et al. 1970
Resonant energy transfer (RET)	Antibody and an analyte derivate are labelled with a donor fluorophore and an acceptor dye respectively. Binding of the tracer to the antibody brings donor and acceptor in close proximity and energy transfer can occur between donor and acceptor.	Van der Meer et al. 1994, Khanna et al. 1980

Proximity assays involve a signal which depends on the distance of two molecular species participating in the assay. The measurable signal depends directly on the distance of a tracer compound and the antibody. Sample, tracer, and antibody can be mixed initially and a readout can be taken after settlement of equilibrium. Some examples are given in Table 1. While the quenching assay (Lin et al. 1997) by definition may be significantly influenced by different compositions of the sample matrix, the FPIA and the RET assay are attractive for environmental screening applications. The FPIA requires only one labelled compound - that is a fluorescence labelled analyte derivate. However the readout is based on the measurement of orthogonally polarised fluorescence signals (Dandliker et al. 1970). This requires defined beam paths, which are not easy to maintain in miniaturised systems. The RET assay in contrast requires two labelled components, but readout is based on fluorescence intensity. Therefore we have chosen the energy transfer assay concept.

1.1.1.5   Resonant energy transfer (RET) immunoassay

The absorption of visible radiation by a chromophore being in the electronic ground state (S0) occurs on a time scale of 10^{-14} s and leaves the chromophore in an excited electronic and vibrational state. The molecule relaxes rapidly into the vibrational ground state of the excited electronic state (S1). Typically this excited state is relaxed by radiationless dissipation of energy. However, if the excited state is long lived (10^{-9} s and more) radiative relaxation becomes probable and fluorescence can be observed.

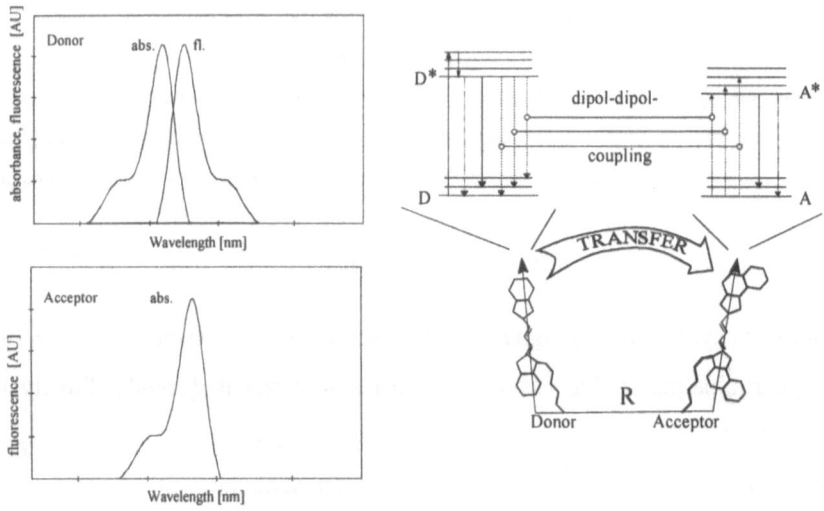

Figure 2:   Principle of resonant energy transfer. Right: the emission spectrum of the donor fluorophore (top) overlaps with the absorption spectrum of the acceptor chromophore (bottom). Right: if both chromophores are in close vicinity, energy from the excited state of the donor can be transferred to the electron system of the acceptor. This is a radiationless process. The rate of RET depends on $1/R^6$

Molecules which emit a significant fraction of absorbed light as fluorescence light are called fluorophores.

RET is a process which occurs from the excited state in concurrence with other processes like fluorescence. RET can occur if a chromophore with an absorption spectrum that overlaps with the emission spectrum of a fluorophore is brought into the close vicinity of the fluorophore (Figure 2). Then the energy from the excited fluorophore (termed the "donor") can be transferred radiationless to the chromophore (termed the "acceptor"). This process, which is due to the electronic interaction of the transition dipole moments of donor and acceptor, leads to quenching of the donor fluorescence. In case the acceptor itself is a fluorophore, a fraction of the transferred radiation can be emitted as acceptor fluorescence (Van der Meeer et al. 1994, Pengguang et al. 1994). The particular attractivity of the RET process is its massive distance dependence.

The efficiency of the RET-process scales with $1/R^6$. The transfer process is characterised by the distance where the probability for RET equals all other processes leading to a relaxation to the ground state. Typical values are around 5 nm. This makes RET an attractive probe for immunoassay applications. RET between a donor and an acceptor attached to an antibody and an analyte derivate respectively will only occur, if the analyte derivate is bound to the antibody. The amount of RET observed, therefore is an indicator of the number of free binding sites (Khanna et al. 1980). RET can be observed either by a decrease in donor fluorescence, or - if the acceptor is fluorescent itself - by an increase in acceptor fluorescence.

1.1.1.6 Quality control and error checking

The very high sample density possible in miniaturised formats allows error checking to be included and controls at the level of each single sample, simply by using more than one cavity per sample. Also the testing of more than one analyte per sample can be easily introduced. We will therefore concentrate on a matrix approach, where each sample is dispensed in multiple cavities within a column of a nanotitreplate. Subsequently different will be dispensed within the rows of the plate (Figure 3). This will allow a limited set of dispensing actions to cover a multitude of analytes with controls and performance checks included at the level of individual samples. The time consuming step in microdrop dispensing is the transfer of the sample to the dispenser unit. Typically this requires repeated flushing of the pipette with the sample. Dispensing itself is a rapid procedure taking one second or less per cavity.

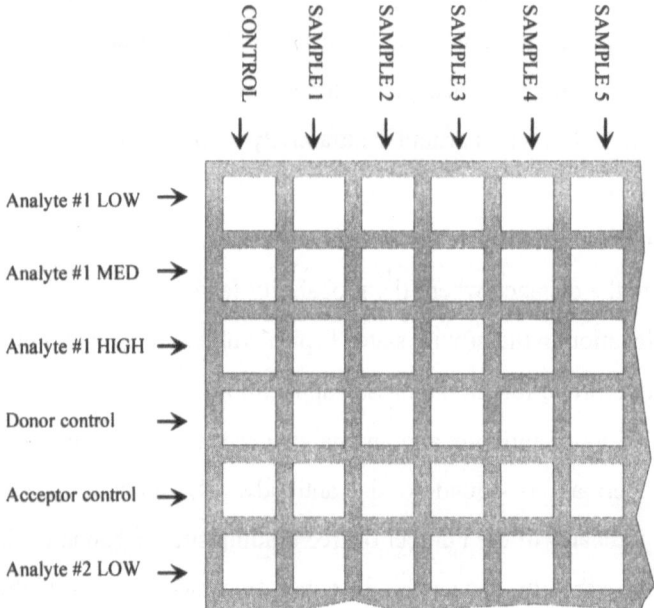

Figure 3: Matrix assay concept. Aliquots of individual samples or control samples are dispensed into multiple cavities of a column of the nanotitreplate. Reagents are dispensed into rows of the plate. The use of multiple wells per sample allows to control assay performance at the level of each sample. The use of a homogeneous phase assay allows adjustment of sensitivity by varying reagent volumes

The use of blank and control samples is common and can be transferred directly to this assay concept. The repeated dispensing of aliquots of the same sample opens new routes to assay design. The accurate dispensing of reagents in small and discrete volumes (individual microdrops) allows the working range of the assay to be shifted by varying the reagent amount. Effects of the sample matrix on the tracer compounds can be identified by dispensing aliqots of the tracer alone.

1.1.2　Materials and methods

Common chemicals and biochemicals were purchased from Sigma Chemical Co., Deisenhofen/Germany and Fluka, Neu-Ulm/Germany. s-Triazine standard solutions were purchased from Riedel de Haën, Seelze/Germany. Cy5-dye, monofunctional reactive NHS-ester and Cy5.5-dye, bisfunctional reactive NHS-ester, were purchased from Amersham Life Science, Braunschweig/Germany. The triazine derivative 4-chloro-6-(isopropylamino)-1,3,5-triazine-2-(6′-amino)caproic acid (atrazine caproic acid, ACA) and the anti-Simazin antibody fragment (Fab), were a gift from Ram Abuknesha, GEC London. Concentration of protein solutions was performed using ultrafiltration-YM-membranes (Microcon-30) obtained from Amicon GmbH, Witten/Germany.

Ultraviolet and visible spectra were recorded on a Zeiss Specord M500 spectrophotometer. The slit was set to 2 nm and the scan speed 60 nm/min.

Fluorescence spectra were recorded on a Perkin-Elmer LS-50 B luminescence spectrometer equipped with a red sensitive photomultiplier R928. The excitation and emission slits were 5 nm, scan speed 120 nm/min and excitation wavelength was 647 nm.

1.1.2.1　Labeling of Fab with Cy5

A solution of 80 μg ($1 \cdot 10^{-7}$ mol) of Cy5 in 30 μl of absolute DMF was added to a solution of 500 μg ($1 \cdot 10^{-8}$ mol) Fab in 250 μl of 0.1 M sodium carbonate-bicarbonate buffer, pH 9.5, and mixed thoroughly by gentle vortexing. The reaction was incubated at room temperature for 30 minutes, with additional mixing approximately every 10 minutes. For separation of unbound fluorescent dye from protein-dye conjugate, the solution was placed into a Microcon-30 sample reservoir and spun at 12000xg for 5 minutes. The residue was washed twice with phosphate buffered saline (PBS), pH 7.4, and the spin repeated until 20 μl remained in the sample reservoir. The Microcon-30 unit was inverted into another vial and the conjugate recovered with a brief spin at 1000xg. The dye/protein ratio was estimated spectro-photometrically as described by Mujumdar et al., 1993. The molar extinction coefficient of Fab and Cy5 were assumed to be 72000 $M^{-1} \cdot cm^{-1}$ at 278 nm, and 250000 $M^{-1} \cdot cm^{-1}$ at 649 nm, respectively. The molar dye:protein ratio of the conjugate was found to be 2:1 with an estimated protein yield of 57%.

1.1.2.2 Labeling of bovine serum albumin (BSA) with Cy5.5

A solution of 170 μg ($1.3 \cdot 10^{-7}$ mol) of Cy5.5 in 30 μl of absolute DMF was added to a solution of 500 μg ($7.2 \cdot 10^{-9}$ mol) BSA in 250 μl of 0.1 M sodium carbonate-bicarbonate buffer, pH 9.5, and mixed thoroughly by gentle vortexing. The reaction conditions and duration, as well as the purification steps were the same as described above. The estimated protein yield was 94%. The dye:protein ratio was determined to be 2:1 ($\varepsilon_{BSA}(278\ nm) = 45540\ M^{-1} \cdot cm^{-1}$ and $\varepsilon_{Cy5.5}(678\ nm) = 230000\ M^{-1} \cdot cm^{-1}$).

1.1.2.3 Preparation of ACA-N-hydroxysuccinimide ester (ACA-NHS)

To a solution of 5 mg ($1.65 \cdot 10^{-5}$ mol) of ACA, and 2.1 mg ($1.8 \cdot 10^{-5}$ mol) of N-hydroxysuccinimide (NHS) in 200 μl of absolute N,N-dimethylformamide (DMF) kept on ice, were added 3.76 mg ($1.8 \cdot 10^{-5}$ mol) of dicyclohexylcarbodiimide (DCC). The mixture was stirred for 6 h at room temperature, filtered, and was used directly (without further purification) for conjugation to proteins.

1.1.2.4   Labeling of BSA-Cy5.5 (1:2) conjugate with ACA-NHS

To a solution of 500 µg ($7 \cdot 10^{-9}$ mol) BSA-Cy5.5 conjugate in 250 µl of 0.1 M sodium carbonate-bicarbonate buffer, pH 9.5, a solution of 15 µg ($5 \cdot 10^{-8}$ mol) of ACA-NHS in 5 µl of absolute DMF was added and mixed thoroughly by gentle vortexing. The reaction conditions and duration, as well as the purification steps were the same as described above. The ACA concentration of the conjugate was determined from absorption spectra. Absorbance contributions from protein and dye were subtracted in order to result in a ACA-spectra. The molar extinction coefficient of ACA was calculated to be 3100 $M^{-1} \cdot cm^{-1}$ at 264 nm. The atrazine:protein ratio was found to be 8.8:1.

1.1.2.5   RET immunoassay

Immunoassays were performed by separately adding various amounts of atrazine and fixed amounts of ACA-BSA-CY5.5 conjugate to solutions containing F_{ab}-Cy5 conjugate in PBS, pH 7.4, containing 100 µg/ml chicken egg albumin (OVA) and 5% methanol. The F_{ab}-Cy5 conjugate concentration and the ACA-BSA-Cy5.5 conjugate concentration in all experiments were 0.1 nM (500 ng/ml) and 0.34 nM, respectively, unless otherwise stated. After each addition of unlabelled atrazine and ACA-BSA-Cy5.5 conjugate, the solution was incubated for 10 minutes at 20°C. After the equilibration, fluorescence data were collected.

1.1.3   Results and discussion

The Amersham Cy5 and Cy5.5 dyes are well suited for excitation with diode laser sources (Figure 4). No significant background fluorescence is expected at these wavelengths. The Stokes shift of both the dyes is sufficient for use in analytical applications. After coupling to protein an about twofold decrease in fluorescence

22

quantum yield was observed. The exact amount depended on the degree of substitution.

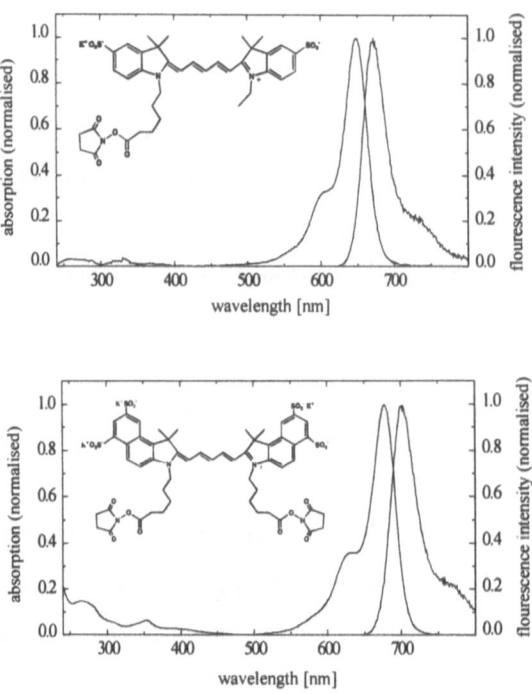

Figure 4:   Absorption and emission spectra of the fluorescent dyes used in this study. Top: Cy5 spectra; bottom: Cy5.5 spectra

The overlap of donor emission spectrum and acceptor absorption spectrum is almost perfect (Figure 5). From the spectral data a Förster distance R_0 of about 7 nm could be calculated. This value is quite high and makes the two dyes a good choice for RET based immunoassays. A slight drawback is the fact that the absorption spectrum of both dyes tails to short wavelengths. This precludes an exclusive excitation of the donor dye by selection of an appropriate wavelength.

The most simple way for construction of a tracer would be to conjugate an analyte derivate directly with the acceptor dye. However, the separation of precursor materials and the conjugate may be quite tedious. Also this straightforward approach allows no adjustment of the dye-to-analyte-derivate ratio. Therefore we have chosen another approach, where bovine serum albumin (BSA) serves as a carrier for tracer construction. This carrier was subsequently labelled by Cy5.5 (acceptor) and an analyte derivate. The overall size of the BSA molecule (about 4 nm) is compatible with the requirements of RET. Also the use of a carrier allows multiple analyte derivates to be linked with a single acceptor molecule. This will reduce acceptor fluorescence due to direct excitation of the acceptor by the laser source. The donor was coupled to anti-Simazine Fab-fragments, to reduce the size of the immune complex and to promote RET.

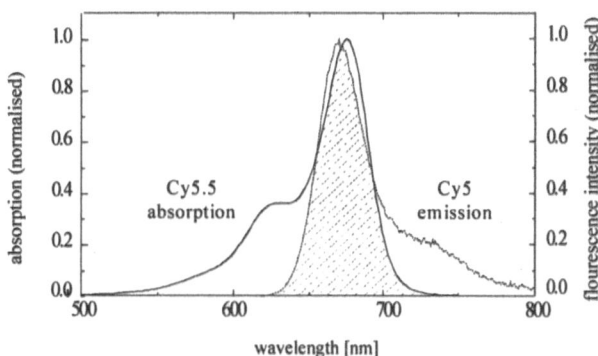

Figure 5: Cy5 (donor) emission spectrum and Cy5.5 (acceptor) absorption spectrum. The overlapping region is shaded. The high degree of overlap of both spectra indicates high efficiency of RET between Cy5 and Cy5.5

For the construction of the assay, the donor-labelled Fab (500 ng/ml) was mixed with increasing amounts of the fluorescence tracer (Figure 6). With increasing concentration of the acceptor-labelled tracer, the donor fluorescence at 670 nm is quenched. A reduction to less than 50% of the initial donor fluorescence was achieved. However, a high excess of tracer bound atrazine with relation to binding sites was required to achieve this amount of quenching. Only little acceptor fluorescence was detected. We presume this is due to a relatively low quantum efficiency of the Cy5.5 acceptor dye. At high amounts of tracer, the Cy5.5 fluorescence becomes more prominent, but mainly due to direct excitation of the acceptor dye by the laser light source. For the construction of a calibration curve, the working concentration of the tracer was fixed to a value where the quenching of the donor was maximum, while the acceptor fluorescence was still limited.

The assay was used to establish a first calibration curve for atrazine (Figure 7). With increasing analyte concentration the tracer is displaced from the antibody and the donor fluorescence at 670 nm increases. The change in acceptor fluores- cence is limited. A calibration curve can be constructed from this by plotting the donor fluorescence vs. the analyte concentration. The calibration curve shows a test midpoint of 30 nM (6ppb). The working range covers about one order of magnitude. The overall signal change from low to high atrazine concentrations is 50% of the initial value. This means a limited degree of modulation of the assay output. The performance at this level is sufficient for semiquantitative testing, but a clear reduction of coefficients of variation is mandatory to give quantitative results.

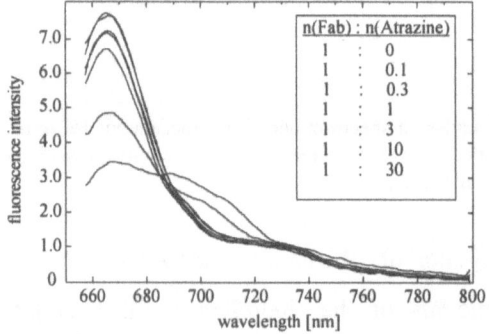

Figure 6: Determination of optimal tracer concentration. Donor-labelled antibody (500 ng/ml) was mixed with increasing amounts of tracer (BSA-Cy5.5-atrazine). The spectra show the decrease of donor fluorescence (at 667nm) and the increase of acceptor fluorescence (700 nm). Part of the acceptor fluorescence is due to direct excitation of the acceptor by the excitation light source. The inset gives the ratio of binding sites and tracer bound atrazine

1.1.4 Outlook

The RET immunoassay works as a wash free, homogeneous phase assay. After mixing of sample and reagents the reaction proceeds towards equilibrium and can be read out without critical timing. The concentrations of antibody and tracer used in the assay must match the target analyte concentrations. For drinking water applications the EU limit is 0.1 ppb which corresponds to concentrations below 1 nM. Assuming an assay volume of 100 nl this corresponds to about 10^7 fluorophore molecules within the sample. In other analytical applications fluorescence detection down to the single molecule level has been demonstrated. Therefore fluorescence detection should be no major obstacle. The RET assay presented gives a test midpoint of 6 ppb. This is not yet sufficient for routine testing in drinking water applications. The competitive assay format used requires high affinity antibodies for trace level determinations. Optimisation of the tracer compound used may also help in shifting the calibration curve to lower concentrations.

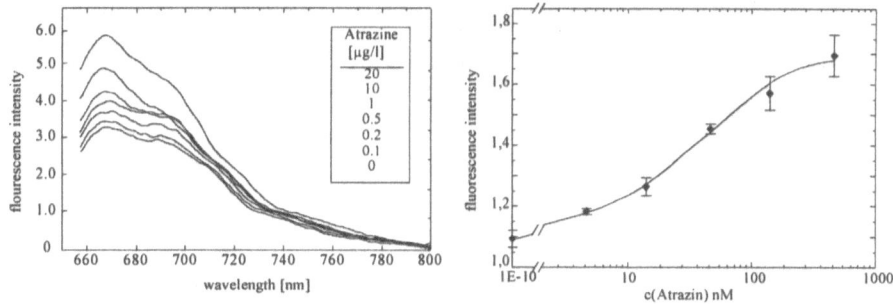

Figure 7: Left: Fluorescence spectra of RET atrazine assay. Increasing amounts of free atrazine displace the tracer from the antibody and lead to increased donor fluorescence. Right: resulting calibration curve. The calibration curve was constructed by plotting the donor fluorescence (at 667nm) vs. the atrazine concentration. A test midpoint of 6 ppb is found. Details are given in text

Practical applications of ink-jet technology in the analytical laboratory are still limited (with the exception of low cost colour printers). In our experience a high content of organic matter may alter and affect the formation of droplets and lead to problems. Also samples will require filtration to be compatible with the handling technique. An increasing use of microdrop techniques in sample handling is expected to improve this situation.

Small amounts of fluid undergo rapid evaporation. This is primarily due to the increased surface/volume ratio of small droplets. The increase in vapour pressure of droplets of 0.1 to 1 mm diameter is minute (Atkins 1990). Therefore, control of the relative humidity of the environment of the nanotitreplate or cooling of the sample carrier close to the dew point can address this problem.

The interrogation of a nanotitreplate with a sample spacing of 0.5 mm to 1 mm is not a major technical problem. Different approaches based on imaging of the entire carrier or sequential interrogation of individual cavities are possible. Dual excitation wavelengths or dual emission wavelengths may be used to improve the amount of information obtained.

All the building blocks for nanotitreplate assays in environmental studies are at hand. The possibility to use generously sample wells and replicates for each sample tested, may offer new approaches to prescreening of environmental samples and to immunoassay strategies itself.

1.1.5   Acknowledgement

This work is supported by the German Ministry of Education, Science, Research and Technology under the "Mikrosystemtechnik 1994 - 1998" programme (project LINDAU / 16SV541 / VDI-VDE IT). The close and productive co-operation with the project partners (Bodenseewerk Perkin Elmer - Überlingen; Bremer Institut für angewandte Strahltechnik BIAS - Bremen; Laser Labor Göttingen - LLG; Institut Dr. Jäger - Tübingen) is gratefully acknowledged.

1.1.6   References

Atkins, P.W. (1990): Physical chemistry. Oxford University Press, Oxford.

Blanchard, A.P., Kaiser, R.J., Hood, L.E. (1996): High-density oligonucleotide arrays. Bios. Bioelect. 11, 687-690.

Dandliker, W.B., de Saussure, V.A. (1970): Fluorescence polarisation in immuno-chemistry. Immunochemistry 7, 799.

Khanna, P.L., Ullmann, F. (1980): 4',5'-Dimethoxy-6-carboxyfluorescein : a novel dipole-dipole coupled fluorescence energy transfer acceptor useful for fluorescence immunoassays. Anal. Bioch. 108, 156-161.

Lin, M., Nielsen, K.J. (1997): Binding of the Brucella abortus lipopolysccharide O-chain fragment to a monoclonal antibody. Quantitative analysis by fluorescence quenching and polarization. Biol. Chem. 272, 2821-2827.

Mujumdar, R. et al. (1993): Cyanine dye labeling reagents: sulfoindocyanine succinimidyl esters. Bioconjugate Chem. 4 (2), 105-111.

Van Der Meer, B.W., Coker III, G., Chen, S.-Y.S. (1994): Resonance Energy Transfer - Theory and Data. Verlag Chemie, Weinheim.

Pengguang, W., Brand, L. (1994): Resonance energy transfer: methods and applications. Anal. Bioch. 218, 1-13.

1.2 Screen-Printed Sensors and Biosensors for Environmental Applications

M. Del Carlo, A. Cagnini, I. Palchetti, S. Hernandez and M. Mascini

Dipartimento di Sanità Pubblica, Epidemiologia e Chimica Analitica Ambientale, Sezione di Chimica Analitica, Università di Firenze, Via G. Capponi 9, 50121 Firenze, Italia

Abstract. In this paper the use of screen-printed electrodes for different environmental applications is reported. Sensors and biosensors can easily be developed varying the modifier of the working electrode surface. The versatility of these devices is demonstrated applying several electrochemical techniques such as amperometry, voltammetry and stripping analysis. Analytical procedures for pesticides, heavy metals and polychlorinated biphenyls are described.

1.2.1   Introduction

The use of the screen printing procedure for the development of sensors and biosensors is gaining consideration in the field of analytical chemistry as shown by the increasing number of scientific papers describing modifications and applications of these tools and the launch of commercial products based on this technology (Cagnini et al. 1997).

Screen printing technology is particularly attractive for the production of disposable sensors (Prudenziati 1994). This is especially important in order to overcome the phenomenon referred to as 'electrode fouling' which is one of the main drawbacks of graphite based electrodes.

With this technique the ink used to print electrodes can be varied easily and therefore different properties of the final sensor can be achieved. The inks can be printed on several kind of supports like glass, ceramic and plastic sheets. Many different inks are commercially available and some of them are based on noble metals (Au, Pt, Ag, etc.). However for these inks a high firing temperature (850-1200°C) is necessary and the overall process becomes cumbersome. In our opinion the most interesting materials for printing electrochemical sensors are the carbon-based inks, because of their very low firing temperature (from room temperature to 120°C) and because they can be printed on plastic sheets. Carbon can also be mixed with different compounds (mediators, enzymes, and metal particles) and therefore modified sensors can be produced. In the literature some carbon electrodes modified with catalytic particles have been described and they showed an improvement in the detection of hydrogen peroxide with respect to simple carbon electrodes. The aim of this work is to show various applications of these devices in environmental monitoring field.

1.2.1.1   Printing of the electrodes

Electrodes were printed with a Model 245 screen-printed obtained from DEK (Weymouth, England), using different inks obtained from Acheson Italiana (Milano, Italy). A graphite based ink (Electrodag 421), a silver ink (Electrodag 477 SS RFU), an insulating ink (Electrodag 970 SS) are used. The substrate was a polyester flexible film (Autostat HT 5) obtained from Autotype Italia. For the applications described below three different devices are used: for the determination of anticholinesterase pesticides a ruthenium-modified carbon screen-printed working electrode with silver screen-printed pseudoreference and counter electrodes (electrochemical strip). For metal determination we used a single round shaped carbon electrode coupled with external reference and auxiliary electrodes. For PCB measurements,however we used the electrochemical strip consisting of the screen-printed carbon working electrode and silver screen-printed pseudoreference and counter electrodes.

1.2.2 Biosensors for pesticide detection

1.2.2.1 Introduction

Organophosphorus and carbamates pesticides are known to be inhibitors of cholinesterase activity and this effect has been monitored through potentiometry, piezoelectric devices, fibre optic systems, amperometry. The use of metal modified screen-printed working electrodes for amperometric measurements with choline oxidase and acetylcholinesterase was investigated in our laboratory. In the proposed method, a ruthenized-carbon screen-printed working electrode was used for hydrogen peroxide detection. Two printed silver layers were used both for the pseudo reference and the auxiliary electrodes. Choline oxidase was immobilized directly on the working electrode surface obtaining a disposable and sensitive choline biosensor. This device was used as detector for the measurement of the inhibition of acetylcholinesterase by pesticides as previously reported (Palchetti et al. 1996) and it was able to detect as little as 10^{-9} M (0.2 ppb) of Carbofuran. Preliminary experiments were performed to monitor pesticides in river water samples. The measurement is based on the following reaction:

$$\text{Acetylcholine} + \text{H}_2\text{O} \xrightarrow{\text{AchE}} \text{Acetate} + \text{Choline}$$

Choline is the substrate of choline oxidase that produces hydrogen peroxide.

Hydrogen peroxide was oxidised on the electrode surface and the current was monitored. Choline oxidase was immobilized on the electrode surface while acetylcholinesterase is free in solution. In this procedure the electrode was immersed in the solution containing 2-4 U/ml of acetylcholinesterase, the baseline was reached and then acetylcholine was added. The current signal has previously been shown to be proportional to the analyte concentration.

1.2.2.2   Materials and methods

Choline oxidase (ChO) from Alcaligenes species (14,6 U/mg of proteins) (EC 1.1.3.17), acetylcholinesterase (AchE) from electric eel type VI S (200-400 U/mg of proteins) (EC 3.1.1.7.), acetylcholine (chloride salt) and choline (chloride salt) were obtained from Sigma Chemicals Co. (St Louis, MO, USA). A standard of carbofuran was purchased from Polyscience Corporation (USA). Standard solutions were prepared daily by dissolving the pesticide in acetonitrile purchased from Sigma (Milan, Italy). Ruthenium (5%) on activated carbon was obtained from Aldrich Chimica (Milano, Italy). All other reagents were analytical grade and were obtained from Merck (Darmstadt, Germany).

A Metrohm 641 VA-Detector for the application of the potential and an Amel Model 337 for the monitoring of current were used. Current was recorded with a Linseis L6512B recorder.

1.2.2.3   Results and discussion

Hydrogen peroxide detection

Amperometric detection of hydrogen peroxide was performed at +700 mV vs pseudo reference silver electrode in 0.1M borate buffer (pH 9.0) with KCl 0.1M. The calibration curve for hydrogen peroxide ($y=0.43x-0.44$, $r^2=0.999$, n=3) shows a detection limit of 10 µM. The coefficient of variation (CV%) of 3 different electrodes at 5 different concentrations of hydrogen peroxide (from 50 µM to 250 µM) was 10,4 %.

Preparation of the choline sensor

Adsorption was used as immobilization procedure for the preparation of the choline biosensor, being a rapid and simple procedure especially for single-use sensors. It was performed by dissolving a fixed amount of choline oxidase (14,6 U/mg) in 0.1 M phosphate buffer pH 7.15 with 0.1 M KCl and then adding it on the working electrode surface. The electrode was left to dry at room temperature and stored dry at 4°C. Five different enzyme concentrations were tested by adding 10 µl of enzyme solution containing 0.087, 0.13, 0.19, 0.26 and 0.39 mg of choline oxidase. For this purpose 0.13 mg of choline in 10 µl of buffer was chosen as the optimized amount. The calibration curves for choline (y=4.18x+0.06; r^2=0.999; n=3) show a detection limit of 50 µM. The coefficient of variation (CV%) for the reproducibility of choline measurements was found to be 12 % testing 3 different electrodes at 4 different choline concentrations (from 0.2 mM to 1.2 mM).

Acetylcholine measurements

In the presence of acetylcholinesterase, acetylcholine is metabolized to choline, according the following reaction:

$$\text{Acetylcholine} + H_2O \xrightarrow{\text{AchE}} \text{Acetate} + \text{Choline}$$

Choline is the substrate of choline oxidase that produces hydrogen peroxide.

Hydrogen peroxide was oxidised on the electrode surface and the current was monitored. Choline oxidase was immobilized on the electrode surface while acetylcholinesterase is free in solution. In this procedure the electrode was immersed in the solution containing 2-4 U/ml of acetylcholinesterase followed by the addition of acetylocholine after reaching a stable baseline. The current signal has previously been shown to be proportional to the analyte concentration.

Measurements of the inhibition

Pesticides are known to inhibit the activity of acetylcholinesterase. In the presence of such compounds, the rate of choline production is reduced with the amount of pesticides. The procedure for obtaining the total anticholinesterase activity using the choline sensor in a pesticide standard solution prepared in buffer was as follows. The standard solution was mixed with the enzyme solution (AchE) reaching a final concentration of 2-4 U/ml and incubated for a constant time. Then the electrode was inserted in the solution and a specific amount of acetylcholine was added (final concentration: 0.5 mM). The current obtained after 1 minute was recorded and it was compared with the current value obtained without pesticides. The calculation of percent of inhibition was obtained according to the following equation:

$$I\% = \frac{I_1 - I_2}{I_1} \times 100$$

where I% is the degree of inhibition, I_1 is the current value in absence of pesticides (blank) and I_2 is the current value in presence of pesticides (sample).

Calibration curves with carbofuran, considered as a reference pesticide, show a limit of detection of $1 \cdot 10^{-9}$ M (Table 2).

Table 2: Inhibition percentage of different carbofuran standard solutions. Each value is the average of three different measurements

Carbofuran (M)	Inhibition (%)	Coefficient of variation (%)
$1 \cdot 10^{-9}$	5.2	20
$1 \cdot 10^{-8}$	31.8	14
$1 \cdot 10^{-7}$	53.4	14
$1 \cdot 10^{-6}$	95.3	1

Pesticides determination in river water samples

The biosensor was used to monitor pesticides concentrations in eight different river water samples. Samples were obtained from Arno river near Florence. Five ml of the sample were filtered (porosity: 0.45 µm) and 190.7 mg of sodium borate and 37,3 mg of potassium chloride were added. The pH was adjusted to 9.0 using concentrated hydrochloric acid. The solution was used for inhibition measurement. For every determination, a blank test was performed using bidistilled water as sample. The inhibition values were then used to calculate Carbofuran equivalent concentrations. These results are to be compared with a standard analytical method (e.g. GC-NPD). The proposed method may suffer from lack of selectivity, though it is the result of the total anticholinesterase activity of a sample. Results are reported on Table 3.

Table 3: Inhibition values obtained from Arno river water samples. Carbofuran concentration values were obtained by interpolating the inhibition values on the reference standard curves

Samples	Inhibition (%)	Carbofuran equivalent (M)
1	37.1	$2\cdot10^{-8}$
2	19.6	$4\cdot10^{-9}$
3	37.3	$2\cdot10^{-8}$
4	26.0	$6\cdot10^{-9}$
5	36.6	$2\cdot10^{-8}$
6	28.9	$8\cdot10^{-9}$
7	25.4	$5\cdot10^{-9}$
8	60.5	$3\cdot10^{-7}$

1.2.3 Screen-printed electrodes for heavy metal detection using anodic stripping voltammetry

1.2.3.1 Introduction

Heavy metals are important environmental pollutants and the knowledge of their real content in various matrices is mandatory. In the last year, decentralised monitoring equipment are more requested. Stripping analysis can be considered the most powerful technique for in field analysis, due to the small size, easy installation requirement and the simultaneous investigation of several metals. Moreover none of the other techniques for heavy metal detection can compete with stripping analysis on the basis of sensitivity per money invested. The sensitivity of stripping analysis is attributed to the preconcentration steps and the advanced measurement procedures. The coupling of screen-printed electrodes with stripping techniques is a revolution in the way of performing stripping analysis (Wang and Tian, 1992), because the design and the operation of stripping system is greatly simplified, in accordance with the requirements of decentralised assays. The aim of this work is to show some results of anodic stripping voltammetry obtained with our devices, using a screen-printed carbon working electrode.

1.2.3.2   Materials and methods

Sodium acetate, acetic acid, nitric and hydrochloric acids were Merck Suprapur. The water used for the preparation of the solutions was supplied from the reagent grade ion exchange system Milli-Q. Mercury(II) chloride was purchased from Merck. Heavy metal stock solutions were prepared by diluting Copper, Lead, Cadmium standard solutions (Fluka). An Amel polarographic analyser Model 433/W (Milano, Italy) was used.

1.2.3.3   Results and discussion

Anodic stripping voltammetry (ASV) was performed as follows. Thin mercury film screen-printed electrodes were prepared before each measurement by electrochemical deposition of mercury. The mercury film was preplated from a nondeaerated, stirred, 80 mg/L mercury(II) solution in HCl 20 mM, by holding the working electrode at the deposition potential (-1 V) for 2 min.

The potential was then switched to -200 mV for 2 min. cleaning period. Subsequent ASV cycles involved the metal deposition and the stripping steps. Experiments were performed in stirred solution and in the presence of dissolved oxygen during the deposition step. The stripping step was performed with a quiescent solution. The potential was scanned using square wave voltammetry. The optimal square wave conditions, for the simultaneous investigation of Cu(II), Pb(II) and Cd(II), were 28 mV amplitude, 3 mV step potential, 15 Hz frequency. Acetate buffer 20 mM at pH 4.7, was chosen as best medium. Each working electrode was used in a disposable manner. Careful disposal and mercury recover has to be considered. The reference electrode was a saturated calomel (S.C.E.), and the counter a platinum electrode (BAS).

Figure 8 shows that on increasing the preconcentration time, the anodic current increases linearly.

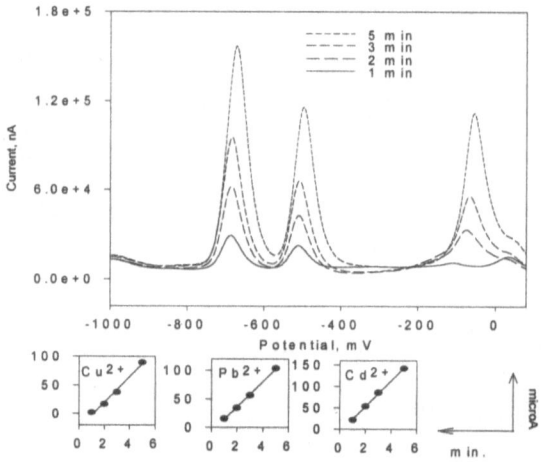

Figure 8: Different preconcentration time: SWASV, 28mV amplit., 15Hz freq.; acetate buffer 20mM, pH4.7

The calibration curves shown in Figure 9 were obtained for a preconcentration time of 3 min. The detection limit, calculated as three times the signal to noise ratio at 10 ppb of the three metals, was under these conditions 0.4 ppb for lead, 1 ppb for cadmium and 8 ppb for copper. These results were obtained with standard solutions. In order to test real samples (e.g. river water, tap water) it will be necessary to evaluate the effect of the various matrix.

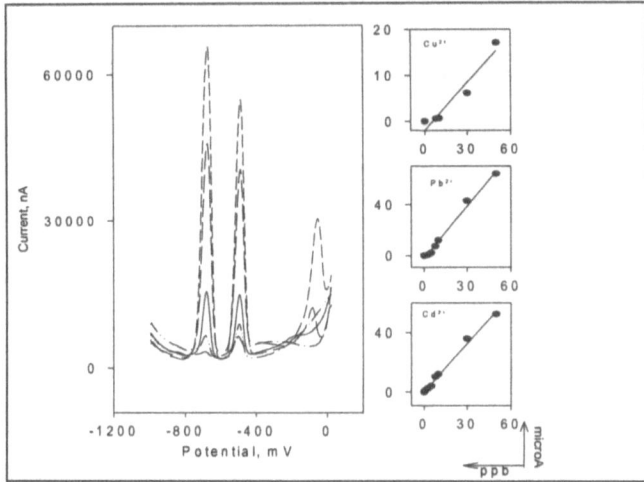

Figure 9: Calibration curve, acetate buffer 20mM pH 4.7

1.2.4   Immunoassay for PCB detection

1.2.4.1   Introduction

Polychlorinated biphenyls (PCB) have been recognised for several years as ubiquitous environmental pollutants. The high toxicity of some of the PCB congeners represent a risk for the public health as these molecules are still present in the environment, even though the production of PCB has been banned (Sanderson et al. 1994). Both during the monitoring and cleaning procedure, PCB analysis is usually carried out using gas chromatography equipped with an electron capture detector, although other detecting strategies (e.g. mass spectrometry) are used. These techniques require expensive instrumentation which are not suitable for on-site analysis, and usually the analysis procedure is time consuming.

Immunochemical detection methods of PCB have been described in the literature varying from radioimmunoassay, to the use of liposome based immunoassay. ELISA is becoming an important tool for the assessment of chemical pollutants. An ELISA for PCB detection with horseradish peroxidase as enzyme label with electrochemical flow injection analysis (FIA) has been recently developed in our laboratory (Del Carlo and Mascini, 1996).

The assay here reported was developed with a commercial apparatus, enzyme linked immunofiltration assay (ELIFA[TM]). The device enables one to perform the assay in 96 flow through wells in which a membrane is placed. The reagents are forced through the membrane by suction and they can be discarded or collected using a microtiter plate.

The assay was performed in a competitive scheme. A bovine serum albumin (BSA) *BSA*-PCB conjugate was the basis for the PCB immobilisation procedure. The immobilisation consisted in the formation of covalent bindings between the preactivated membrane and the BSA. After the competition for the antibodies binding sites, the amount of anti-PCB IgG that reacted with the immobilised PCB was evaluated using a secondary, alkaline phosphatase labelled, antibody. The detection of alkaline phoshatase was made with α–naphtyl phosphate and the screen-printed electrodes with differential pulse voltammetry.

1.2.4.2   Materials and Methods

Carbonate and bicarbonate sodium salts, and methanol were from Merck, Germany. Polyoxyethylene-sorbitanmonolaurate (Tween 20), ethanolamine (EA), diethanol-amine (DEA), bovine serum albumin (BSA), α–naphtol, α–naphtyl phosphate, and alkaline-phosphatase(AP)-labelled antichicken IgG were from Sigma, USA.

PCB-BSA molecules have been prepared as previously described (Del Carlo and Mascini 1996) and the anti-PCB antibodies, have been purchased by Bioreclamation Inc. (New York, USA). The Easy-Titer[TM] ELIFA[TM]System was from Pierce, USA. The preactivated membrane was an Immobilon[TM] AV Affinity membrane from Millipore (Bedford, MA, USA). The polystyrene microtiter plate were purchased from Corning NY, USA. The peristaltic pump was a Gilson model Minipuls (Villiers les Bel,

France). The Aroclor 1260 standard was obtained from Polytechne (Livorno, Italia). Bidistilled water was obtained with a MilliQ system, Millipore Inc.

All the electrochemical measurements were performed using a polarographic analyser computer controlled model 433 A (Amel, Milano, Italia).

Differential pulse voltammetry (DPV) measurements

The measurements were carried out by depositing 100 µl of α–naphthol standard solution on the electrodes device, after a fixed time the voltage scan was performed (i.e. 10 seconds for the α–naphtol calibration curve and for the assay assessment, and 2 minutes for the enzyme activity evaluation).

The simplex method was used for the optimisation of two parameters (the scan speed and pulse amplitude) considering as final goal the highest sensitivity. An initial simplex of three vertices was considered with a fixed search step. All the DPV experiments were performed in the potential range 0/+400 mV vs Ag screen-printed pseudoreference electrode, with a pulse width of 50 milliseconds.

The optimisation was performed using a 50 µM α–naphtol standard solution in DEA pH 9.6.

Alkaline phosphatase detection

In order to evaluate the sensitivity of alkaline phosphatase detection using screen-printed electrodes, the enzyme hydrolysis of α–naphtyl phosphate was evaluated using the DPV. A stock solution of alkaline phosphatase (10 U/ml) was prepared using DEA buffer. Via serial decade dilutions a number of alkaline phosphatase standards were obtained in the range $1-1 \times 10^{-4}$ U/ml.

An aliquot of 90 µl of 1mg/ml α–naphtyl phosphate solution was deposited on the screen-printed electrodes device. Afterwards 10 µl of enzyme solution was added and the solution was homogeneized by pipetting. After 2 minutes the DPV measurement was performed. Three replicates for each standard were carried out.

Enzyme linked immunofiltration (ELIFA[TM]) assay for PCB

In this work an enzyme linked immunofiltration assay (ELIFA[TM]) device was used, allowing the binding of proteins to a membrane. The ELIFA[TM] working unit is composed of three pieces of precision cut plexiglass (sample application plate, transfer cannula holder and collection chamber) which are designed with accompanying gaskets to seal tightly thus eliminating cross-talk and providing constant flow rate from well to well. ELISA principles apply to the ELIFA[TM] except that the latter takes advantage of filtering the initial ligand solution through a membrane to bind the protein conjugate to it, whereas using polysterene microtiter plates a 16 hours incubation is usually required. ELIFA[TM] system facilitates binding of high levels of ligand compared with the polystyrene surface of microtiter plates. Moreover it should override the effects of limiting diffusion of the reagent to the solid phase.

The solution containing the proteins to be immobilised on the membrane are forced through the membrane layer by suction when the vacuum is applied using a peristaltic pump.

The competitive assay was performed on a preactivated membrane, where covalent binding occurs.

First of all the collection chamber and the membrane support plate have to be clamped together. A 8x12 cm square of Immobilon[TM] membrane piece was placed on top of the silicone gaskets. After reagents addition, the vacuum was produced by the peristaltic pump. The membrane was wetted with carbonate buffer solution and the coating solution (20 μg/ml of BSA-PCB in the same buffer) was flowed through the membrane for 15 minutes. After a washing step, the blocking solution (500 μl/well) was sucked through the membrane in 2 hours time. Another washing step followed. Then, different standards of PCB (Aroclor 1260), in the range 10 ng/ml-10μg/ml, were incubated for 30 minutes with a limiting amount of IgG anti PCB (1μg/ml) in PBS. Afterwards 200 μl of this solution was added for the competition with the immobilised BSA-PCB which was completed in 20 minutes, followed by a washing step. The alkaline phosphatase labelled anti chicken IgG antibodies solution (1:750 dilution) was sucked through in 20 minutes. The last washing was performed before the substrate solution (1 mg/ml α–naphtyl phosphate) addition. Finally, the substrate was flowed through the membrane in 15 minutes. The product obtained from the enzymatic reaction was collected in microtiter plate. Each well contained 10 μl of phosphate solution 0.1 M as stop solution. α–Naphtol (100 μl) was then pipetted onto the screen-printed device for the DPV measurement.

1.2.4.3   Results

Differential pulse voltammetry measurements and optimisation

Within the voltammetric techniques, DPV is one of the most sensitive voltammetric techniques. The most effective parameters on the current response in this technique are commonly considered the scan rate and the pulse amplitude. The simplex trials and observations are reported in Table 4. The first three runs define the initial simplex.

Table 4: Current responses using DPV during the simplex search procedure

	X_A PULSE AMPLITUDE mV	X_B SCAN RATE mV sec^{-1}	Y RESPONSE μA	
1	10	20	1.41	*Original simplex*
2	10	40	1.91	*Original simplex*
3	30	30	5.81	*Original simplex*
4	40	50	7.25	
5	50	40	8.28	
6	50	60	9.30	
7	70	50	9.87	
8	70	70	11.9	
9	90	60	9.3	*Reject simplex*
10	50	80	8.4	*Reject simplex*
11	70	70	11.8	

The search procedure led to define the simplex with 6-7-8 vertices. Point 8 was retained for three successive simplexes and response at this point was determined again (vertex 11). As it remained the highest response, it was considered as the optimum value with the simplexes of the chosen size. Smaller simplex could not be assessed due to instrumental limitations (i.e. Amel 433 in the range 10-100 mV sec^{-1} only allows 10 mV sec^{-1} steps). Therefore a scan rate of 70 mV sec^{-1} and a pulse amplitude of 70 mV were used in the following measurements.

The detection of naphthol with screen-printed electrodes and DPV showed a great sensitivity (252 nA/µM), good reproducibility (4.6% and 3.3% for naphthol 10^{-6} and 10^{-4} M respectively, n=7), wide measurable range (5.0×10^{-7}-10^{-4} M), and ease and speed of use. The voltage range of the measurement was from 0 mV to 400 mV vs the Ag screen-printed pseudoreference electrode. Using a scan rate of 70 mV sec^{-1}, a measuring time of 5.5 seconds was obtained. This is faster than a flow injection analysis response time, so increasing the number of samples analysis with respect to this widely used technique. The reported calibration curve (y=252x+900, r^2=0.998) allowed a calculated detection limit of 0.1 µM. The excellent reproducibility of the method combined with the speed of the measurement, the non-hydrodynamic working conditions, the low detection limit and the simplicity of the apparatus required made DPV the technique of use for the enzyme activity and immunoassay assessment.

Alkaline phosphatase measurements

A calibration curve of alkaline phosphatase was obtained in the range 1-1x10^{-5} U/ml. The standard error evaluated for 5 replicates of 1x10^{-2} U/ml was 6.8% using 2 minutes of enzyme substrate incubation. The detection limit was 2.1x10^{-6} U/ml (calculated using three times the noise of naphthyl phosphate solution without enzyme at the peak potential) and was comparable to that obtained using other substrates and analogous incubation time and various techniques described in the literature.

Enzyme linked immunofiltration (ELIFA^{TM}) assay for PCB

The immunoassay for Aroclor 1260 was performed with the ELIFA system, resulting in a fast binding of the BSA-PCB to the solid phase (30 minutes vs 16 hours using polysterene plates). The assay was suitable to measure standards in the range 0.01-10 µg/ml, considering a 10% decrease of the blank signal as the detection limit. This assay range makes it suitable to be used with soil, sludge and waste water samples. The main drawback of this assay is the high background signal. This is possibly due to cross reaction of the anti-PCB IgG with the BSA molecule used as immobilising molecule, infact as reported by the producer, the immunizing molecule was a BSA conjugate. The highest Bx/Bo% was in fact 50% of the blank signal (for the 10 µg/ml standard) thus limiting the sensitivity of the assay. A change of the conjugating molecule was envisaged to lead to wider dynamic range combined with higher sensitivity.

1.2.5 Conclusions

This overview shows the versatility of screen-printed electrodes technology for environmental applications.

Coupling the transducer with biological components (e.g choline oxidase and acetylcholineesterase) makes this sensors an interesting in-field device for the detection of organophosphorous and carbamate pesticides. The use of the bare graphite electrode is suitable for trace metals detection using ASV. The sensors have been also used as detection system for an immunoassay for PCB. The use of disposable devices well replace other analytical methods that can not be used under field conditions. The proposed methods are also attractive with respect to traditional methods for their low cost and ease of use.

1.2.6 References

Cagnini, A., Palchetti, I., Del Carlo, M., Mascini, M. (1997): Thick film sensors and Biosensors. In: Conference Proceedings Vol. 54 (Sbeveglieri, Tondello, eds.). SAA96 Bologna.

Del Carlo, M., Mascini, M. (1996): Enzyme immunoassay with amperometric flow injection analysis using horseradish peroxidase as a label. Application to the determination of polychlorinated biphenyls. Anal. Chim. Acta 336, 167-174.

Palchetti, I. Cagnini, A., Del Carlo, M., Coppi, C., Mascini, M., Turner, A.P.F. (1997): Determination of anticholinesterase pesticides in real samples using a disposable biosensor. Anal. Chim. Acta. 337, 315-321.

Prudenziati, M. (1994): Thick film sensors. Vol 1, Handbook of Sensors and Actuators. Elsevier.

Sanderson, T.J., Noestrim, R.J., Elliot, J.E., Bellward, G.D. (1994): Biological effects of polychlorinated dibenzo-p-dioxins, dibenzofuran, and biphenyls in double crested cormorant chicks. J. Toxicol. Env. Health 41 (2), 247-265.

Wang, J., Tian, B. (1992): Screen-printed Stripping voltammetric/potentiometric electrodes for decentralized testing of trace lead. Anal. Chem. 64, 1706-1709.

1.3 A Manufacturing Technology for Biosensing

Johanna C. Morey

Bookham Technology Ltd, Rutherford Appleton Laboratory, Chilton, Oxfordshire OX11 0QX, UK

Abstract. There is a current emphasis on sensors for environmental monitoring and biological applications, as reflected in the scientific literature. In particular, optical waveguides as evanescent wave sensors are sensitive analytic probes giving direct information about surface binding reactions. Miniaturised, integrated sensors offer benefits for real-time monitoring at low cost. However, there is still a gap to be bridged between research activity and the commercial exploitation of biosensors in terms of the manufacture and cost of sensing devices. Bookham Technology Ltd. have developed a process for the production of Active Silicon integrated Optical Circuits, ASOC™, which is based upon techniques commonly employed in the semi-conductor industry to provide a cost-effective, high-volume manufacturing solution. With sources and detectors integrated with optical waveguide interferometers, ASOC™ has the potential to provide a multiple analysis system on a single chip.

1.3.1   Introduction

Biosensing is very much an active research area with potentially large markets in environmental monitoring, medical diagnostics and process control for liquid or gas-based systems. However, relatively few laboratory-demonstrated systems have broken into the market as commercial products. The reasons for this include sensor performance, particularly regarding the ease and reproducibility of measurement, and manufacturing reliability and cost.

The increasingly prominent field of integrated optics offers a viable manufacturing route for optical waveguide-based sensors for biosensing applications. Optical circuits, i.e. interconnecting waveguides for guiding and directing light, can be fabricated on small chips. Evanescent wave sensing is a direct, real-time measurement technique for affinity reactions and involves the interaction of the evanescent field of a waveguide mode with the cover layer above the waveguide. The coupling of evanescent wave sensing with the use of integrated optical interferometers results in a sensitive and selective sensing platform.

Silicon is an inexpensive base material for the mass production of integrated optical components. Using techniques from the semi-conductor industry, Bookham Technology Ltd. have developed a manufacturing technology for Active Silicon integrated Optical Circuits (ASOC^{TM}) to produce high performance optical components for the telecommunications and sensor markets. This technology allows the fabrication of low loss optical waveguides and the integration of laser diodes and photodetectors. The active capability of the waveguides allows phase modulation for signal interrogation schemes.

1.3.2   An overview of the ASOC^{TM} technology

ASOC^{TM} is designed for the manufacture of integrated optical components, such as transceivers, switches and interferometric devices for fibre optic applications and optical system designs. The technology allows manufacture at low cost and in high volumes. Low loss silicon ridge waveguides are fabricated on a silicon-on-insulator (SOI) wafer with a silicon oxide cladding (Rickman et al. 1992, Rickman 1994). The waveguides are designed for single mode propagation. A silicon substrate with a pattern of optical waveguides can interconnect laser diodes and photodetectors, as depicted in Figure 10 for a transceiver product.

The active capability of the waveguides means that the phase of the guided light can be modulated. The waveguides are doped so that upon current injection the refractive index of the guide changes, resulting in a change in the phase of the guided light. This is important for switching in telecommunications systems, but also has applications for the interrogation of sensing systems. Figure 11 shows active Mach Zehnder interferometers on an SOI chip for use in signal processing systems.

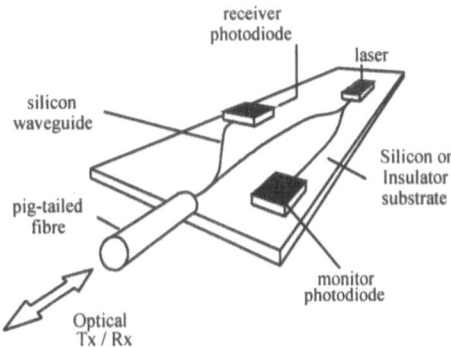

Figure 10: An ASOC^{TM} transceiver

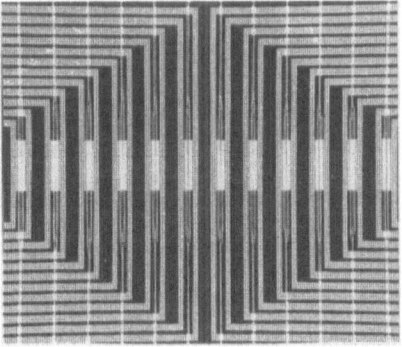

Figure 11: Active Mach Zehnder interferometers fabricated using ASOC^{TM}

The use of phase modulation allows the tracking of the quadrature point in interferometric sensing systems giving increased accuracy. Alternative methods to phase modulation by carrier injection include wavelength modulation of the source, which may be subject to noise and variations in temperature, and thermal modulation of the effective index in one arm of the interferometer. This latter method may be difficult to implement in an integrated optical device due to a temperature induced phase shift being inadvertently applied to both arms instead of one.

1.3.3 On-chip evanescent wave sensing using Mach Zehnder interferometry

The reported use of evanescent wave sensing in conjunction with integrated optical interferometers is becoming more prevalent. The use of ion-exchange waveguides in glass with application for the detection of triazines has been described (Gauglitz and Ingenhoff 1994), and an interferometer within a nitride slab waveguide structure for monitoring immunoreactions has been presented (Heidemann et al. 1993). The fabrication of a GaAs/AlGaAs-based integrated optical interferometer for refractometry has been reported (Maisenhölder et al. 1997).

For evanescent wave sensing using an integrated optics device, part of the optical waveguide structure has the upper cladding removed to form a sensing region or „window". Within the window a surface coating is deposited containing a specific molecular binding reagent for a target molecule. The binding reaction results in a change in refractive index of the surface layer that is detected by the interaction of the evanescent field of the guided wave with this layer. This approach results in a generic sensor platform with various possible chemical and biological sensing applications. The sensor concept may be extended to produce a multiple sensing array with different receptor reagents for various target species.

Evanescent wave sensing in conjuction with interferometry offers a sensitive and selective detection method. A Mach Zehnder interferometer containing a sensing and a reference arm formed from optical waveguides is illustrated in Figure 12. The waveguide upper cladding is removed to form a window in the sensing arm, leaving the higher index waveguide ridge exposed. The surface within the window is deposited with a biological receptor layer. As the target molecule binds to the receptor layer there is a change in the refractive index of the waveguide surface and hence a change in the effective index of the guide.

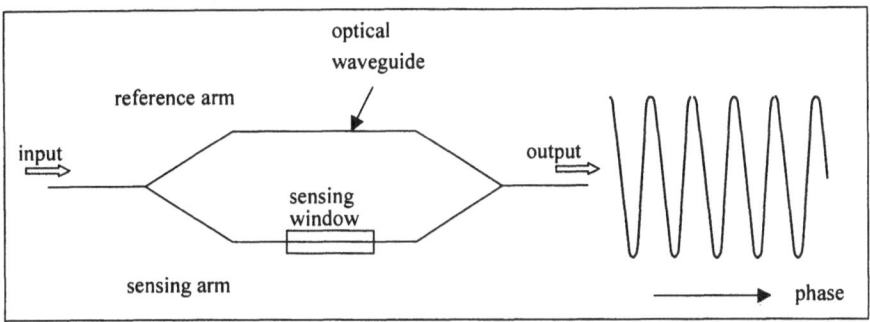

Figure 12: Schematic of an integrated optical Mach Zehnder interferometer

The resulting change in optical path difference between the two arms means the light in the sensing arm undergoes a phase shift with respect to that in the reference arm. This phase change is directly related to the change in the effective index of the guide from the clad to the exposed waveguide, along with the length of the sensing window. Interference fringes are observed at the output of the device, with the intensity (I) related to the maximum intensity (I_o) and the phase change ($\Delta\phi$) as follows:

$$I = \frac{1}{2} I_o [1 + \cos\Delta\phi] \qquad (1)$$

By monitoring the intensity change with time, the phase information can be related back to the concentration and thickness of the bound layer. The use of a reference arm compensates for temperature changes.

Practical considerations involve the integration of the sensing interferometer within a signal interrogation system. The materials used for fabrication of the sensing interferometer waveguide structure should be compatible with the manufacturing technology, and also have refractive indices such that the interaction of the evanescent field of the waveguide mode with the target results in a sufficient „sensing" signal. A waveguide structure with a silicon nitride core and a silicon dioxide cladding fits both these criteria and allows a removable sensing chip to be coupled to an interrogating ASOC^{TM} chip. The sensing chip can then be a disposable or semi-disposable device; the ability to reuse a sensing chip will depend upon the reversibility of the surface chemistry. With ASOC^{TM} phase modulation techniques, resolution to a thousandth of a fringe can be obtained giving sub-ppm changes in the refractive index of the cover layer.

1.3.4   Conclusion

A manufacturing technique has been developed that addresses some of the foreseen problems for the commercialisation of biosensors. The ASOC^{TM} technology is a reproducible method for the volume manufacture of optical sensors at low cost. The combination of evanescent sensing with on-chip interferometry gives a highly sensitive measurement technique with the ability for rapid analysis and continuous monitoring. The sensor can be made specific to molecules of interest depending upon the surface treatment. There is also the potential for multiple analysis as an array of sensor devices can be fabricated on the same chip.

1.3.5   References

Rickman, A.G., Reed, G.T., Weiss, B.L., Namavar, F. (1992): Low loss planar optical waveguides fabricated in SIMOX material. IEEE Photonics Tech. Lett. 4 (6), 633-5.

Rickman, A.G. (1994): Silicon integrated optics and sensor applications. Sensor Review 14 (1), 27-29.

Gauglitz, G., Ingenhoff, J. (1994): Design of new integrated optical substrates for immuno-analytical applications. Fres. J. Anal. Chem. <u>349</u>, 355-359.

Heideman, R.G., Kooyman, R.P.H. and Greve, J. (1993): Performance of a highly sensitive optical waveguide Mach-Zehnder interferometer immunosensor. Sens. & Act. B <u>10</u>, 209-217.

Maisenhölder, B., Zappe, H.P., Moser, M., Riel, P., Kunz, R.E., Edlinger, J. (1997): Monolithically integrated optical interferometer for refractometry. Electron. Lett. <u>33</u> (11), 986-988.

1.4 Development of an Enzyme-Linked Immuno-sorbent Assay for the Determination of Irgarol 1051

M.-P. Marco[1], B. Ballesteros[1], I. Ferrer[2], J. Casas[3] and D. Barceló[2]

Departments of Biological Organic Chemistry[1] and Environmental Chemistry[2]. Animal Breeding Facility[3]. C.I.D.-C.S.I.C., Jorge Girona, 18-26, 08034-Barcelona, Spain

Abstract. An enzyme-linked immunosorbent assay (ELISA) has been developed for the antifouling agent Irgarol 1051. Three different immunizing haptens have been used to obtain antisera (As) with different selectivities. Hapten disseny has been studied according to molecular mechanic considerations (MM2+) and correlated to the avidities of the obtained As. The ability of Irgarol 1051 to compete for the antibody (Ab) binding sites with eleven horseradish peroxidase enzyme tracers (HRP tracers) differing on the chemical structure of the hapten has been investigated. Several usable assays have been obtained showing good sensitivities. One of them has been optimized to obtain a reproducible immunoassay with the following features: IC_{50}= 0.074 µg/l, Slope=1.68, Working range= 0.121-0.036 µg/l and LOD=0.020 µg/l. The immunoassay has been validated with spiked and real seawater samples by high-performance liquid chromatography with diode array detection (HPLC-DAD).

1.4.1 Introduction

Pilot survey studies of coastal waters contamination has been mainly focused on tributyltin (TBT), however its use has been restricted after the regulations introduced by the European Community and the Mediterranean region (Europeenne 1989, UNEP

1989). The herbicide Irgarol 1051 (2-methylthio-4-*tert*-butylamine-6-cyclopropyl-amine-*s*-triazine) has been the substitute of TBT and is mainly used in antifouling paints as a biocide agent in combination with copper- and zinc-based agents. As other triazines Irgarol 1051 is predominantly present in the dissolved phase. However because of its low solubility in water (Irgarol 1051, 7 mg/l; atrazine 33 mg/l; terbutryn 25 mg/l)) and high partition coefficient (Tolosa et al. 1996) (Irgarol 1051 Koc=3.0; atrazine Koc=2.1; terbutryn Koc 2.8) one can predict a higher affinity for the particulate matter compared to other triazine compounds. Not many studies have been performed regarding toxicity effects of this compound, although it has been reported that the photosynthetic activity of the periphyton was significantly lowered at concentrations ranging from 0.25 nM to 1 nM (0.063 to 0.25 µg/l) under long-term studies. Besides the specific activity of the triazines herbicides, it is also worth noting the reported toxicity in the rainbow trout assay where Irgarol 1051 showed an EC_{50} of 0.86 mg/l, while for atrazine the value is ten time higher, 8.8 mg/l (Toth et al. 1996).

Before year 1992 contamination of coastal waters by Irgarol 1051 was not known. First report appeared on year 1993 indicating the presence of this agent in the coastline of Côte d'Azur (Readman et al. 1993) at concentrations ranging from 0.11 to 1.7 µg/l. One year later coastal water contamination by Irgarol 1051 on Southern England was also reported (Gough et al. 1994) at slightly lower (0.002 to 0.5 µg/l), but still significant concentrations. Finally last year, presence of Irgarol 1051 was noted contaminating other regions in Europe such as the Western coast of Sweden (Dahl and Blanck 1996) and in the freshwater of lake Genève (Toth et al. 1996).

Methods of analysis for Irgarol 1051 determination are common to other triazines (gas chromatography (GC) with nitrogen and phosphorous (NPD) or mass spectrometry (MS) detection and HPLC-DAD or MS) reaching very low detection limits (5 ng/l) by preconcentrating high sample volumes (0.5 to 1l). Among other benefits, immunoassays are able to reach very low detection limits without the need to introduce a clean-up or preconcentration step, increasing throughput analysis. The

54

work presented here for the first time deals with the development of a sensitive immunoassay for Irgarol 1051 determination. The assay has been validated and used to carry a one year monitoring study of the contamination by this antifouling agent of the Masnou marina (North of Barcelona, Spain).

1.4.2   Hapten design

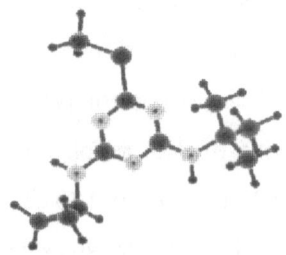

Figure 13: MM2+ model steroview of Irgarol 1051

The hapten should preserve as much as possible the chemical structure, spatial conformation and electronic distribution of the target analyte analyte. Irgarol 1051 is a triazine compound with two voluminous alkyl substituents (cyclopropyl and tert butyl). Additionally, the third position of the triazine ring is occuped by a methylthio group that may confer special properties to the molecule because of the electronegativity and the size of the sulfur atom. The MM2+ optimized geometry of Irgarol 1051 (see model steroview in Figure 13) shows the cyclopropyl group placed perpendicularly outside of the plane defined by the triazine ring. Similarly, the tertbutyl group overpass perpendicularly the above mentioned plane in all directions by the three voluminous methyl groups.

Since all three groups seemed to be potential antigenic determinants, we decided to study the influence of these three groups on the antibody recognition. With this purpose, we prepared by classical synthetic organic methods three immunizing haptens (**4c**, **4d** and **4e**), each of them preserving at least two of these substituents (see Figure 14).

1.4.3   Preparation of the antisera

The immunogens were keyhole limpet hemocyanin (KLH) conjugates. Each of them was used to immunize three female white New Zealand rabbits (**4c-KLH**, R13, R14 and R15; **4d-KLH**, R16, R17 and R18; **4e-KLH**, R19, R20 and R21). The immunization protocol took about six months and involved boosting the rabbits monthly with 100 μg of the immunizing conjugates emulsified with Freund's adjuvant. Small blood samples were obtained from the ear central arteria, ten days after each boost and used to measure antibody titer. When an acceptable titer was obtained the animals were exanguinated by canulation of the carotida. Blood was collected in vacutainers provided of serum separating gel.

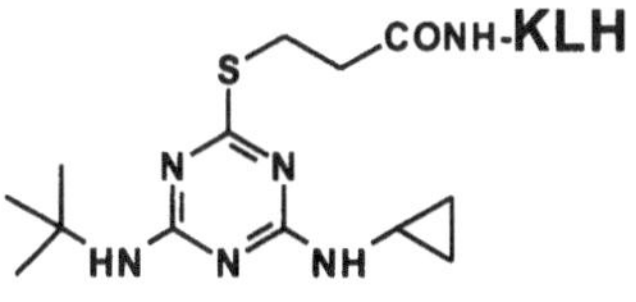

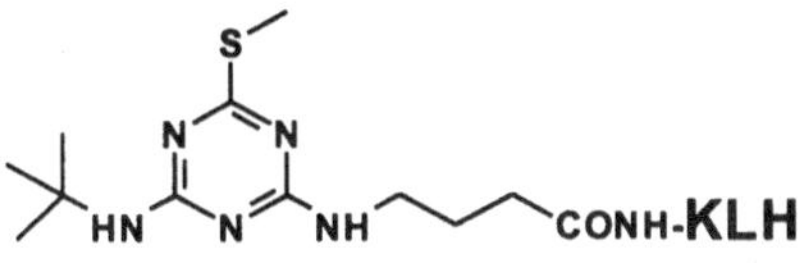

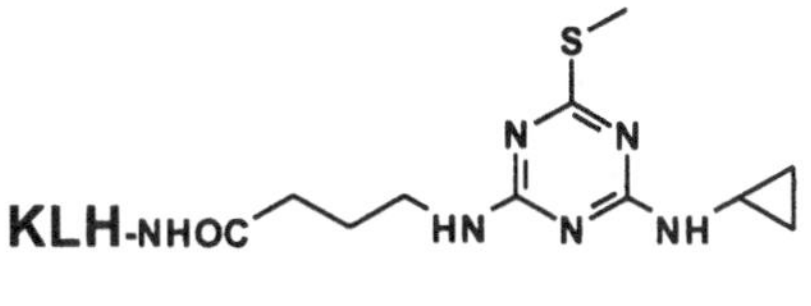

Figure 14: Chemical structures of the immunizing haptens prepared to study the influence of the different substituents of the triazine ring on the antibody recognition

1.4.4 Direct competitive immunoassays

A direct competitive immunoassay is characterized by the ability of the analyte to displace the formation of the HRP tracer-Ab complex. Two equilibria reactions take place simultaneously leading to the analyte-Ab (Ka) and the HRP tracer-Ab ($K'a$) immunocomplexes. The relative value of these two affinity constants is the key to obtain a good and sensitive immunoassay. To our knowledge, nobody has carried out yet an exhaustive investigation on the optimun value of the $Ka/K'a$ ratio. These kind of studies are not trivial, but would provide us of helpful information for the design of appropriate competitors. Therefore, we screened a battery of eleven HRP tracers with the aim to find the hapten giving the most sensitive immunoassay and to correlate the chemical structure of the haptens used as competitors with the IC_{50}s obtained.

1.4.4.1 Influence of the immunizing hapten chemical structure

After screening for competitive assays we found that haptens **4c** and **4d** gave by far the best antisera. Both of them provide several competitive immunoassays showing most of them acceptable IC_{50}s for direct environmental analysis of pesticides. In contrast, the immunizing hapten **4e** did not rend during the screening step any usable assay. Only **As19** was able to afford some competitive assays with IC_{50}s close to 1 µg/l, but it

Table 5: Competitive immunoassays showing IC_{50}s lower than 1 µg/l

Immunogen	As	No. Assays $IC_{50} < 1\,µg/l$
4c-KLH	13	11
	14	11
	15	4
4d-KLH	16	7
	17	6
	18	8
4e-KLH	19	0
	20	0
	21	0

was still insufficient for direct environmental analysis. Table 5 shows the number of assays obtained with each immunogen. showing IC_{50}s values lower than 1 µg/l.

At the light of these results we tried to find an explanation to the low quality encountered for **4eKLH** as immunizing hapten. If the MM2+ optimized geometries of Irgarol 1051 and hapten **4e** are considered, we can explain this by the noncovalent interactions that participate in the stabilization of the analyte-Ab complex: i) lack of *hydrophobic interactions*: the absence of the bulky *tert*-butyl group in the chemical structure of **4e** has lead to antibodies that do no establish hydrophobic interactions with this important hydrophobic area of the Irgarol 1051 molecule; ii) lack of *a hydrogen bond*: the amino group of the spacer arm of the hapten 4e may had produced antibodies that tend to establish a hydrogen bond with the analyte at this point; however, the amino group of the Irgarol 1051 is not accessible because the shielding effect caused by the tert butyl group; and iii) lack of *electrostatic interactions*: the high density of hydrogen atoms of the tert butyl group creates a large area with a slight positive charge in the Irgarol 1051 molecule and simultaneously the negative charge of the triazine ring may be polarized in this direction. Therefore the Abs raised against immunizing haptens possessing the tert butyl group are able to establish more efficient interactions with the analyte, yielding as a consequence better immunoassays.

1.4.4.2   Influence of the chemical structures of the HRP tracers

As it was mentioned before the sensitivity of the immunoassay is highly dependent on the ratio of the affinity constants participating in the equilibria. Looking at the IC_{50}s obtained for a single immunogen and the eleven HRP tracers screened, we assume that for low avidity As the possibility to obtain a sensitive assay will be higher if the chemical structure of the enzyme tracer differs to a high extent from that of the analyte (see behavior of As14 on Table 6). In contrast, for As with high avidity for the analyte

Table 6: Competitive immunoassays obtained with As13, As14 and As15 (**4c**-KLH) and **2a-2f,4a-4e**-HRP tracers

HRP tracer	R1	R2	R3	As IC50 (µg/l)				
				<0.1	0.1...0.3	0.3..1	1..10	>10
2a	Cl	ethyl	(CH2)3COOH		13-14-15			
2b	Cl	isopropyl	(CH2)3COOH		13-14-15			
2c	Cl	ethyl	(CH2)5COOH		13-	15	14	
2d	Cl	isopropyl	(CH2)5COOH		13-15	14		
2e	Cl	terbutyl	(CH2)3COOH			13-15	14	
2f	Cl	cyclopropyl	(CH2)3COOH		13-15		14	
4a	S(CH2)2COOH	ethyl	ethyl			13-15	14	
4b	(CH2)3COOH	isopropyl	ethyl			13-15	14	
4c	(CH2)3COOH	tertbutyl	cyclopropyl		15	13	14	
4d	SCH3	tertbutyl	(CH2)3COOH	15		13-14		
4e	SCH3	cyclopropyl	(CH2)3COOH	15		13	14	

it seems possible to obtain good assays even under homologous or quasi-homologous conditions, because this allows for higher dilution of the immunoreactants (see behavior of As 15 on Table 6).

1.4.5 Immunoassay features and validation by HPLC-DAD

As15/4eHRP was selected from all competitive immunoassays for further optimization and application to the analysis of spiked and real environmental samples.

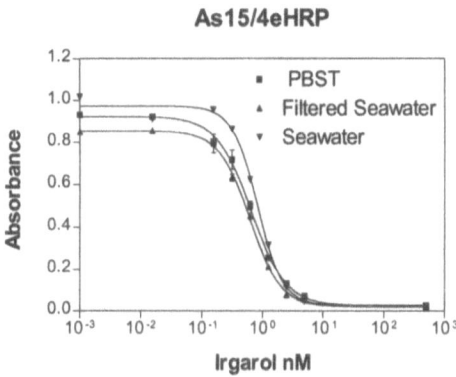

Figure 15: Effect of the seawater matrix in the Irgarol 1051 immunoassay. The small effect caused by the seawater can be avoided by filtering the water prior to ELISA analysis

Parameters optimized were pH, ionic strength and incubation times. The optimized assay shows an IC_{50} of 0.074 ± 0.033 µg/l $(n=9)$, a slope of 1.68 ± 0.26 $(n=9)$, a working range between 0.121 and 0.032 µg/l and a limit of detection of 0.020 µg/l.

With this assay we studied, on a first instance, the effect of the matrix on the immunoassay performance. As it is shown on Figure 15, we could appreciate that seawater does not affect significantly the immunoassay although filtering of the water may lead to more accurate results. In contrast to most other matrices, measurements

60

carried out directly on that matrix may underestimate Irgarol 1051 in environmental samples because of the slightly larger absorbance observed at low concentration values (see Figure 15).

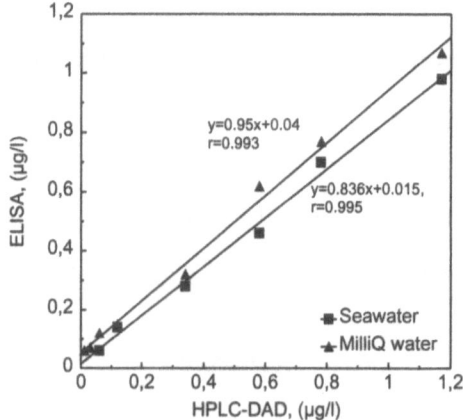

Figure 16: Correlation observed between measurements made by ELISA and HPLC-DAD using spiked milliQ and seawater samples

The cross-reactivity studies yielded information on the selectivity of the immunoassay As15/4eHRP. It was again evident, the importance of the tert-butyl group for the antibody recognition. These triazine herbicides possessing a methylthio group in their structure are weakly recognized in this system (i.e. prometryne 15%), while these having a tert-butyl group are more detected (i.e. terbutylazine 29%) indicating that this group constitutes an stronger antigenic determinant. The best recognized triazines are those containing both groups (i.e. terbutryn 78 %).

We studied accuracy of the assay using water samples spiked at different levels (0.2 to 1.5 nM). In this experiment, the recoveries obtained were always close to 100 %. Similarly, we compared the optimized ELISA with HPLC-DAD by measuring spiked milliQ and seawater samples. The correlation obtained was always good (r> 0.99), and the slope of the linear regression equations was close to 1 for the milliQ water and slower when measurements were performed on seawater samples which is in agreement with the matrix effect previously mentioned (see Figure 16).

1.4.6 Application of the Irgarol 1051 immunoassay to the analysis of environmental samples

We carried out a monitoring survey of the north coast of Barcelona (Masnou marina) for Irgarol 1051 contamination, starting on April 96. Samples were collected monthly

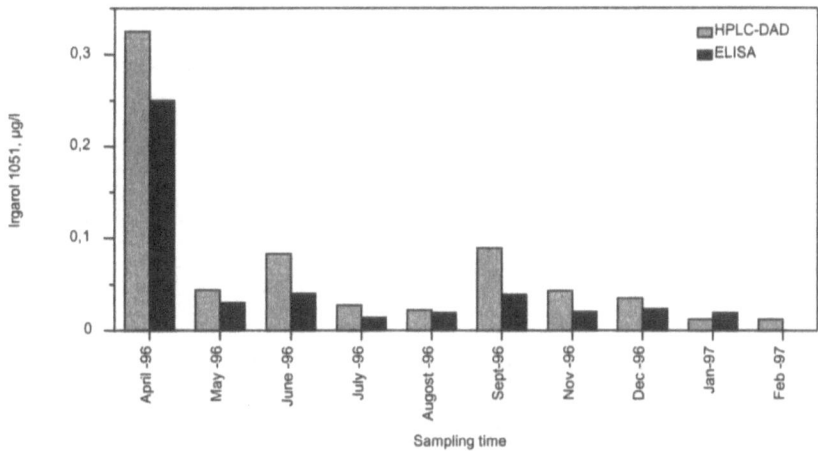

Figure 17: Results of the monitoring study carried out monthly at Masnou marina (Barcelona, Spain). Levels of Irgarol 1051 reached a peak on April when the high boating season starts

at 1 m depth from the surface layer and stored on precleaned amber glass bottles at 4°C until analysis was performed. The study was carried out by ELISA and *on line* SPE-HPLC-DAD and the positive samples were confronted by *on line* SPE-HPLC-APCI-MS. The levels of Irgarol ranged between 0.1 and 0.325 µg/l, similarly to those reported in Southern England; (Gough et al. 1994) and slightly lower than those of Cote d'Azur (Tolosa et al. 1996). As it can be seen in Figure 17, the concentration of Irgarol 1051 reached a peak on April when the high boating season starts. Similar results have been reported by other authors (Dahl and Blanck 1996, Toth et al. 1996).

1.4.7   Conclusions

We have presented in this paper the development and validation of a highly sensitive immunoassay to determine the antifouling agent Irgarol 1051 in seawater samples. The proposed immunochemical technique constitutes an excellent approach for screening this kind of samples. Although the information available on the toxicity of Irgarol 1051 is very limited it is worth noting that the levels reported in this study, as well as those encountered in other coastal waters (western coast of Sweden, Cote d'Azur, southern England) are close to those reported to have significant effects on the periphyton photosynthetic activity (Dahl and Blanck 1996). Therefore, screennig techniques to monitor antifouling agents, such as the reported ELISA, are necessary and, as it has been demonstrated here, may constitute useful tools to assess coastal water contamination.

1.4.8   Acknowledgements

This work has been supported by the EEC Environment Program (Contract No. ENV4-CT95-0066) and PLANICYT (AMB96-2808-CE). We thank Dr. Oriol Bulvena and M-Angeles García (Laboratorios Viñas, Barcelona, Spain) for their kind assistance in animal treatments and Ciba-Geygy (Barcelona, Spain) for supplying samples of simazine, atrazine and Irgarol 1051.

1.4.9   References

Dahl, B., Blanck, H. (1996): Toxic Effects of the Antifouling Agent Irgarol 1051 on Periphyton Communities in Coastal Water Microcosms. Mar. Pollut. Bull. 32, 342-350.

Europeenne, C. (1989): Directive du Conseil, pp. 19-23. European Community, Journal officiel des Communautes Europeennes.

Gough, M.A., Fothergill, J., Hendrie, J.D. (1994): A survey of southern england coastal waters for the s-triazine antifouling compound Irgarol 1051. Mar. Pollut. Bull. 28, 613-620.

Readman, J.W., Liong Wee Kwong, L., Gronding, D., Bartocci, J., Villeneuve, J.-P., Mee, L.D. (1993): Coastal Water Contamination from a Triazine Herbicide Used in Antifouling Paints. Environ. Sci. Technol. 27, 1940-1942.

Tolosa, I., Readman, J.W., Blaevoet, A., Ghilini, S., Bartocci, J., Horvat, M. (1996): Contamination of Mediterranean (Cote d'Azur) Coastal Waters by Organotins and Irgarol 1051 used in antifouling paints. Mar. Pollut. Bull. 32, 335-341.

Toth, S., Becker-van Slooten, K., Spack, L., de Alencastro, L.F., Tarradellas, J. (1996): Irgarol 1051, an Antifouling Compound in Freshwater , Sediment, and Biota of Lake Geneva. Bull. Environ. Contam. Toxicol. 57, 426-433.

UNEP (1989): "Report of the sixth ordinary meeting of the contracting parties to the convention for the protecction of the Mediterranean sea against pollution and its related protocols". UNEP(OCA)/MED IG. 1/5.

1.5 Immunosensor Systems with Renewable Sensing Surfaces

M. Santandreu, S. Solé, S. Alegret and E. Martínez-Fàbregas

Grup de Sensors & Biosensors, Departament de Química, Universitat Autònoma de Barcelona, 08193 Bellaterra, Catalonia, Spain

Abstract. A complication emerges when antigens and antibodies interact in continuous-use immunosensor systems. This complication comprises the regeneration of the biological sensing surface. In the present work we report the development and the study of two strategies designed to overcome this limitation.

The first strategy is based on the construction of amperometric immunosensors using rigid immunocomposites. These materials contain a conducting polymer composite that acts as a support for the bulk-immobilized immunological material. The surface of these immunosensors is renewable. A simple polishing procedure uncovers a fresh immunocomposite surface ready for a new immunoassay. This contrasts with conventional, single-use devices. Furthermore, immunosensors of different sizes and shapes can be produced using these immunocomposites. The closeness between the immunoconjugate enzyme-label and the conducting sites on the surface of the sensor yields a higher electron transfer efficiency. This is clearly convenient when building amperometric devices. The simplicity of this strategy makes it particularly convenient for manual immunoassay methodologies.

The second strategy is based on an immunochemical analysis system featuring flow injection techniques. This system uses potentiometric detection with immunochemical reagents immobilized on magnetic particles where the sensing surface can be renewed after each analysis. Measurements are reproducible since the magnetic particles can be fixed to the surface of the sensor at will. The regeneration of the sensing surface is achieved by turning on or off a magnetic field. This is especially convenient in flow systems where other approaches to surface renewal may be difficult or cumbersome. The simplicity and flexibility of this strategy makes it particularly convenient for automated immunoassay methodologies. It is also versatile because a wide choice of immunological reagents can be used.

These two immunosensor systems were applied to the measurement of RIgG using a competitive technique. They were also used in the detection of GaRIgG using a sandwich technique, where peroxidase was the enzyme label for amperometric measurements and urease the label for potentiometric measurements.

1.5.1   Introduction

Immunoassay techniques are becoming increasingly important in the clinical field (Owen 1994, Blanchard et al. 1990, Heinemann and Halsall 1985) for determination of drugs (Blake and Gould 1984, Wang 1988), metabolites, steroids and hormones (Duan and Meyerhoff 1994). At present, environmental applications of these analytical techniques are spreading (Blanchard et al. 1990, Nelson et al. 1995, Marco et al. 1995), to analyze pollutants such as pesticides (Hammock et al. 1980, Kindervater et al. 1990, Gascon et al. 1995, Kaláb and Skládal 1995, Dzantiev and Zherdev 1996), industrial waste materials (Blanchard et al. 1990) and degradation products.

Heterogeneous immunoassay techniques require a solid phase with immobilized immunoreagents. This configuration allows the separation of the free immunoreagent and the bound immunoconjugate after a wash cycle.

Several techniques for the immobilization of the immunological material have been described (Hall 1990). These techniques include the immobilization of the immunoreagent on different polymer surfaces and activated membranes (Glazier and Rechnitz 1991, Sansubrino and Mascini 1994), capillary tubes (Rogers et al. 1991),

66

other tubing (Rook and Cameron 1981), controlled pore glasses (CPG) (Lee and Meyerhoff 1988), gels (Shellum and Gubitz 1989), optic fibers (Wijesuriya 1994), etc. However, it is important that the capacity of the immunoreagent to combine with its complementary is not impaired by the immobilization process (Blake and Gould 1984).

An immunosensor where the immunological material is immobilized on a transducer is an example of the heterogeneous techniques mentioned (Figure 18). The main advantage of immunosensors over classic immunoassays is the proximity between the biological material and the transducer. This closeness makes the immunosensor more sensitive to the antibody-antigen interaction. Unlike classic immunoassay techniques, immunosensors generate a measurable signal directly, they detect the antigen quickly and they are potentially reversible (Miller 1981, North 1985, Lee and Meyerhoff 1988, Wijesuriya 1994, Leech 1994, Byfield and Abuknesha 1994, Killard et al. 1995). Additionally, immunosensors are more robust methodologies than immunoassays since the immunological material is on the transducer, and more economical since they require less reagent to operate.

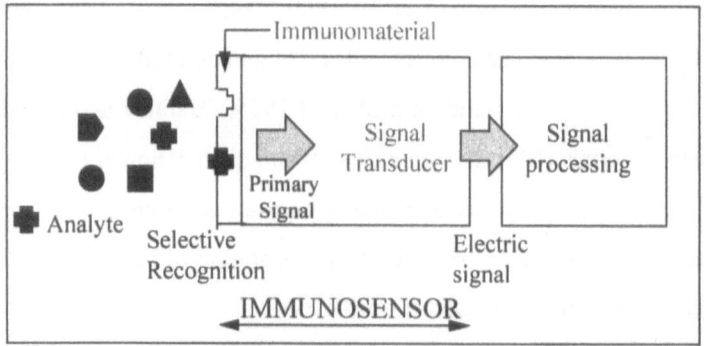

Figure 18: Schematic representation of the functioning of an electrochemical immunosensor

No covalent bonds are present in antigen-antibody interactions, but multiple weak bonds (van der Waals forces, hydrogen bridges, hydrophobic interactions, etc.), that produce a high association constant (Marco et al. 1995). Once it is bound to its conjugate, the immunoreagent on the transducer is not available for further reactions unless the established bond is broken. This difficult regeneration of the sensing surface limits the application of the otherwise advantageous immunosensor (Miller 1981, Buerk 1993), especially in continuous-use and automated applications.

Two regeneration strategies developed in our laboratories are described in this chapter, namely, amperometric immunosensors based on rigid immunocomposites (Santandreu et al. 1997) and immunosensors formed by magnetic immunoparticles adapted to a flow injection system.

Rabbit and goat immunoglobulins (RIgG and GaRIgG, respectively) were used as the immunological model to evaluate both immunosensors.

In a further study, these immunosensors will be used to the measurements of PCBs.

1.5.2   Immunocomposites

Immunocomposites are a novel alternative to the immobilization of immunological materials for the construction of amperometric immunosensors.

An immunocomposite results from the combination of two or more components where one of them is the immunological material. Each phase keeps its individual features while the resulting material has new physical, chemical and biological characteristics (Alegret 1996, Céspedes and Alegret 1996). Table 7 shows the advantages of using rigid conducting immunocomposites.

Amperometric biosensors based on rigid polymer matrices, graphite powder and an enzyme (biocomposites) constitute a precedent of this type of device (Alegret et al. 1996). Several biosensors of this kind have been developed in our laboratories (Céspedes et al. 1993a, Céspedes et al. 1993b, Martorell et al. 1994, Alegret 1996, Alegret et al. 1996a, Alegret et al. 1996b, Morales et al. 1996, Martorell et al. 1997).

Table 7: Key features of the conducting rigid immunocomposites used in the construction of amperometric immunosensors

---

SIMPLICITY

The preparation of the immunocomposites is simple and is carried out using dry chemistry tecniques. The resulting immunosensors are inexpensive and can be considered for single-use applications.

MOULDABILITY AND RIGIDITY

The immunocomposite is highly mouldable before curing, permitting the construction of amperometric immunosensors of different shapes and sizes. The construction procedure is compatible with thick film technology. The immunocomposites are characterized by their mechanical stability and rigidity after curing. The surface can then be polished or mechanically altered.

CONTROLLED ACTIVE SURFACE

The components on the sensing surface can be controlled by adjusting their content in the bulk of the immunocomposite.

REPRODUCIBILITY AND REGENERATION

Each new surface yields reproducible results if all the individual components are homogeneously dispersed in the bulk of the immunocomposite. The surface of the immunosensors can be regenerated by simply polishing, obtaining fresh immunocomposite ready to be used in a new immunoassay.

BIOLOGICAL COMPATIBILITY

The immunocomposite acts as a reservoir for the biological material.

ENHANCED ELECTROCHEMICAL RESPONSE

The morphology, dimensions and distribution of the conducting particles allow a microelectrode array behaviour (efficient mass transport, high signal to noise ratio, fast response times, low detection limits).

The polymeric matrix is the insulating phase of the composite and is formed by a resin (epoxy, methacrylate, polyurethane, silicone, etc.) and a hardener. This phase is responsible for the physical, chemical and biological stability of the composite. Graphite is the conducting phase and is responsible of a quality electrochemical response (Alegret 1996).

One of the attractive aspects of using immunocomposites in sensor devices is the rigidity they acquire after curing. This is helpful to overcome one of the limiting features of immunoassay techniques: the regeneration of the active surface. The reuse of the biocomposites can be attained by eliminating the surface where the immunological reaction has occurred by a simple polishing procedure. The polishing produces a new surface with fresh immunological material usable for a new assay.

1.5.2.1   Immunocomposite preparation and biosensor construction

The preparation of the immunocomposite is quick and simple. The ingredients are added in the prescribed quantities and they are mixed thoroughly to produce an homogeneous mixture.

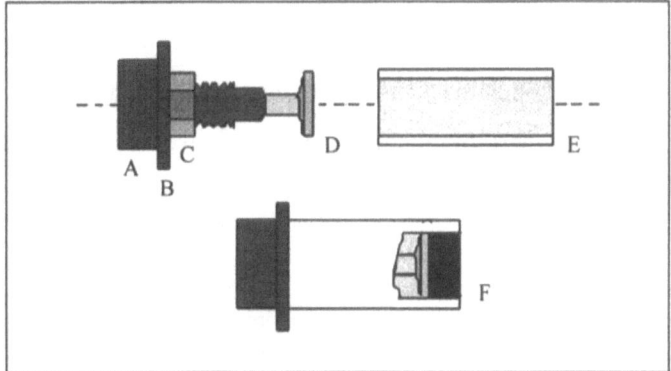

Figure 19:   Construction of the amperometric immunosensor. (A) female connector formed by a 2 mm gold-plated beryllium-copper contact and a polyimide molding, (B) bakelite washer, (C) metal nut, (D) circular conducting copper piece soldered to the connector end, (E) PVC tubing (6 mm i.d., 18 mm length), (F) final aspect of the supporting assembly with a 3 mm thick cup where the immunocomposite sits

Two immunocomposites were evaluated:

1. RIgG-epoxy-graphite is prepared by mixing graphite powder and epoxy resin in a 1:4 (w/w) ratio. Once this paste is homogenized, RIgG is added until its presence in the mixture is 0.9 % in weight. The resulting paste is introduced into a 6 mm i.d.

PVC tube until it makes contact with an electrical connector that sits on the other end of the tube. The composite is cured for a week at 40 °C.

2. The RIgG-methacrylate-graphite is prepared similarly with a graphite to methacrylate ratio of 1:1 (w/w). The immunocomposite is cured in a nitrogen atmosphere at room temperature for three days.

When the immunosensors were not in use they were stored at 5 °C.

Figure 19 shows the components of the immunosensors built.

1.5.2.2   Immunocomposite materials used

Table 8 shows the experimental conditions of the immunoassays used to compare the performance of immunosensors built with either one of the immunocomposites. Alkaline phosphatase was used as the label of the immunoconjugate species.

As shown in Figure 20, RIgG calibration curves for both immunocomposites show the same trend, although the sensitivity (slope) for the immunosensors based on methacrylate is higher. This immunocomposite has two additional advantageous features. First, it is easier to manipulate than the epoxy-based material.

Table 8: Experimental conditions for the determination of RIgG (competitive assay) and GaRIgG (sandwich assay) using amperometric immunosensors based on conducting rigid immunocomposites and enzyme-labeled immunoconjugates. The experimental conditions are listed on Table 9

Immunocomposite	Graphite-methacrylate with 0.9 % RIgG
	Graphite-epoxy with 0.9 % RIgG
Immunoconjugate label	ALKALINE PHOSPHATASE
Immunoassay technique	Competitive

Table 8, continued

Surface pretreatment	Sonication (2 minutes)
Immunoconjugate (stock solutions) Blocking buffer	GaRIgG alkaline phosphatase conjugate (commercial solution: 800 enzymatic units/ml) 0.99 % (v/v) in blocking buffer. BSA 1 % (w/v) prepared in 0.1 M Tris-HCl, 0.001 M EDTA buffer at pH 7.0
First incubation	Definite volumes of a stock RIgG solution were mixed with a fixed volume of the solution of GaRIgG alkaline phosphatase conjugate and enough 0.1% (w/v) BSA solution to give a final volume of 1800 µl Time: 30 minutes
Second incubation	The immunosensor surface is introduced in the solution resulting from the first incubation Time: 2 hours
Incubation conditions	Room temperature Constant stirring
Rinsing step	A 0.1 M Tris, 0.1 M KCl pH 7.5 buffer
Regeneration	Polishing with abrasive and alumina papers

Table 8, continued

Immunocomposite	Graphite-methacrylate with 0.9 % RIgG	Graphite-methacrylate with 0.9 % RIgG
Immunoconjugate label	PEROXIDASE	
Immunoassay technique	Competitive	Sandwich
Surface pretreatment	Sonication (2 minutes)	Sonication (2 minutes)
Immunoconjugate (stock solutions) Blocking buffer	GaRIgG peroxidase conjugate (commercial solution: 100 enzymatic units/ml) 0.99 % (v/v) in blocking buffer. BSA 1 % (w/v) prepared in 0.1 M phosphate buffer at pH 7.0	RaGIgG peroxidase conjugate (commercial solution) 0.99 % (v/v) in blocking buffer. BSA 1 % (w/v) prepared in 0.1 M phosphate buffer at pH 7.0
First incubation	Definite volumes of a stock RIgG solution were mixed with a fixed volume of the solution of GaRIgG peroxidase conjugate and enough 0.1% (w/v) BSA solution to give a final volume of 90 μl. Time: 30 minutes	Definite volumes of a stock GaRIgG solution were mixed to enough 0.1% (w/v) BSA solution to give a final volume of 45 μl. This volume was placed on the immunosensor surface. Time: 30 minutes

Table 8, continued

Second incubation	A 45 μl drop of the previous solution was placed in the surface of the immunosensor. Time: 30 minutes	A 45 μl drop of the RaGIgG peroxidase conjugate stock solution was placed in the surface of the immunosensor. Time: 30 minutes
Incubation conditions	Room temperature	Room temperature
Rinsing step	A 0.1 M disodium phosphate, 0.1 M KCl pH 7.0 buffer	A 0.1 M disodium phosphate, 0.1 M KCl pH 7.0 buffer
Regeneration	Polishing with abrasive and alumina papers	Polishing with abrasive and alumina papers

Additionally, it was experimentally demonstrated that the epoxy-based immunocomposite is less reproducible than the one containing methacrylate. RSD for each measurement using a methacrylate based immunosensor is similar to the 7% described previously (Santandreu et al. 1997). This may respond to the higher porosity of the epoxy resin since a higher porosity means a larger surface that favors the access of antigens.

Table 9: Experimental conditions of the electrochemical measurements used in the immunological determinations represented in Figures 1.3, 1.4 and 1.7. * Amperometric measurements were carried out with a dialysis membrane fixed on the sensing surface

Label	Detection system	Mediator	Substrate	Working media
ALKALINE PHOSPHATASE	Amperometry* $E_{applied}$ = 800 mV (vs SCE)	none	Phenyl phosphate	Tris-HCl 0.1 M, KCl 0.1 M, pH = 7.5
PEROXIDASE	Amperometry* $E_{applied}$ = -100 mV (vs SCE)	Hydroquinone	Hydrogen peroxide	phosphate 0.1 M, KCl 0.1 M, hydroquinone $1.8 \cdot 10^{-3}$ M, pH = 7.0
UREASE	pH ISFET (potentiometry) I_{DS}=100μA; V_{DS}=500 mV	none	Urea	NH_4Cl 0.1 M

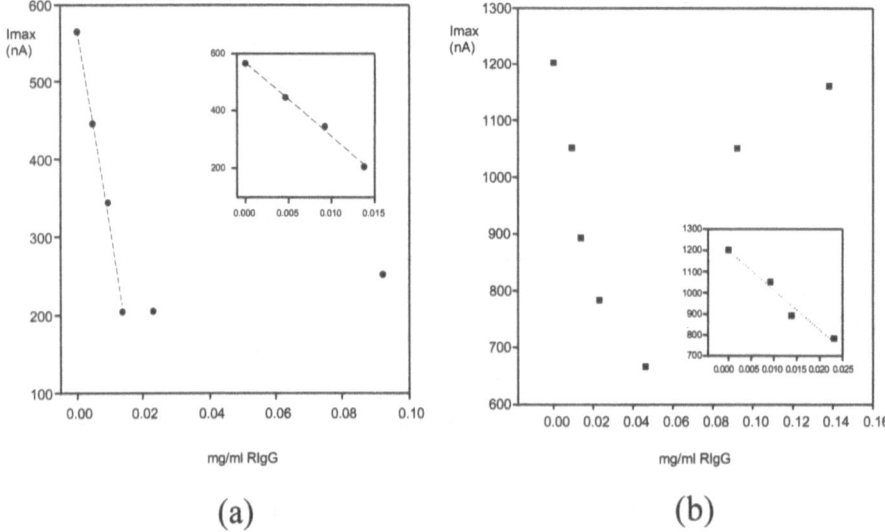

Figure 20: Calibration curves for the determination of RIgG. Competitive immunoassay using alkaline phosphatase as the label of the immunoconjugate. Immunosensors based on immunocomposites: a) RIgG-graphite-methacrylate (Regression: slope = -2572.8 ± 491.9 nA·ml/mg; y-intercept = 568 ± 42 nA; r = 0.998) and b) RIgG-graphite-epoxy (Regression: slope = -1875.1 ± 1042.9 nA·ml/mg; y-intercept = 1199 ± 148 nA; r = 0.98). I_{max} values were obtained by adjustment of experimental data from the phenyl phosphate calibration to Michaelis-Menten equation (Santandreu et al. 1997). Further experimental details on Tables 8 and 9

However, the larger surface also favors nonspecific adsorptions when there is a high content of immunological material in the incubation solution, while this adsorption was not detected experimentally in the methacrylate immunosensors (Santandreu et al. 1997). The higher porosity also means a greater difficulty in achieving reproducible surfaces.

1.5.2.3 Labels: alkaline phosphatase and peroxidase

A competitive technique has been applied to the evaluation of two enzyme labels, alkaline phosphatase and peroxidase. Table 8 shows the experimental conditions used in these assays. Table 9 shows the electrochemical detection parameters for the enzyme labels used. There were two factors that favored the use of peroxidase. The substrate of the enzyme (hydrogen peroxide) is more stable than the substrate of alkaline phosphatase (phenyl phosphate). This is true both for their commercial form and for the calibration solutions. The amperometric response when peroxidase is used is higher and more reproducible. Currents in the µA range as opposed to nA currents when alkaline phosphatase is employed as the label are achieved.

1.5.2.4 Sandwich and competitive techniques

Two different strategies for the use of the immunocomposites have been evaluated. The RIgG composite was evaluated for RIgG determination using a competitive technique and for GaRIgG using the sandwich technique. The experimental conditions for both techniques are shown in Table 8. In both cases peroxidase was the enzyme label. Results are shown in Figure 21.

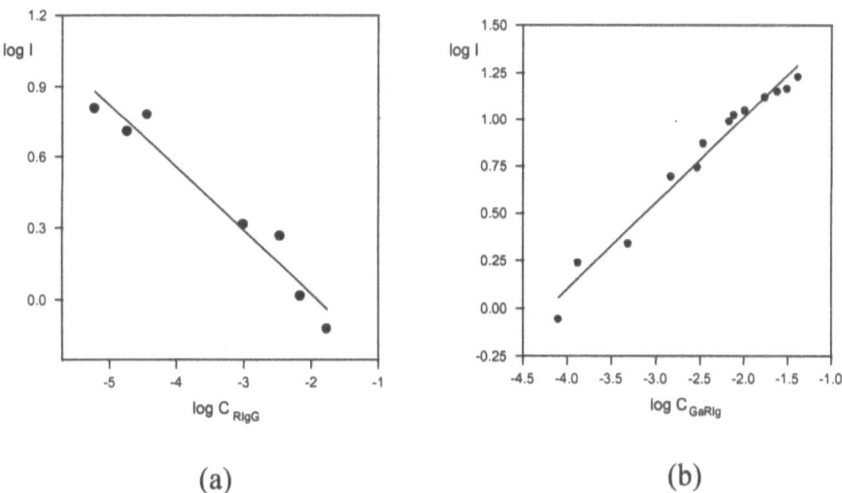

(a)   (b)

Figure 21: (a) Calibration curve for the determination of RIgG (competitive assay), (Regression: slope = -0.27 ± 0.07; y-intercept = -0.50 ± 0.25; r = 0.97) and (b) Calibration curve of GaRIgG determination (sandwich assay), (Regression: slope = 0.43 ± 0.04; y-intercept = 1.88 ± 0.10; r = 0.990). Immunosensor based on a RIgG-graphite-methacrylate immunocomposite. Peroxidase was used as the label for the immunoconjugate. Intensity values were obtained after adding $8.68 \cdot 10^{-4}$ M H_2O_2. Further experimental details in Tables 8 and 9

When plotting log I versus log $C_{analyte}$ a linear portion was present up to 0.04 mg/ml GaRIgG ($2.6 \cdot 10^{-7}$ M) using the sandwich technique. The linear range for the competitive technique was up to 0.02 mg/ml RIgG. The detection limits, defined as the concentration that yields a signal three times the background noise, were $9 \cdot 10^{-12}$ mg/ml GaRIgG ($5.62 \cdot 10^{-17}$ M) and $2 \cdot 10^{-9}$ mg/ml RIgG ($1 \cdot 10^{-14}$ M) respectively.

The reproducibility of the signal for each concentration was higher with the sandwich technique. This responds to the excess of labeled immunological material used in this technique. This means that small variations in the added quantity of the immunological material does not modify the immunological interaction notably.

1.5.2.5   Renovation of the immunosensor surface

The rigidity of the immunocomposites facilitates the mechanical modification of the sensor surface. The regeneration of the solid phase where the immunological material is immobilized can be achieved by polishing it with abrasive paper. After this initial polishing the surface is polished again with alumina paper to ensure a uniform and smooth surface.

In this way the immunological material is renewed without using abrupt changes of pH or ionic force which are the strategies mentioned in the literature for the regeneration of the immunological bonding.

The RIgG immunosensor was incubated repeatedly in a solution of GaRIgG peroxidase conjugate 0.63 % (v/v) in BSA. After the corresponding surface polishing between assays, the amperometric signal showed a RSD of 7 % (n=7) using a peroxide concentration of $4.7 \cdot 10^{-4}$ M, as shown in Figure 22. This proves that polishing the surface of the immunosensor produces a new surface with fresh immunological material capable of interacting reproducibly.

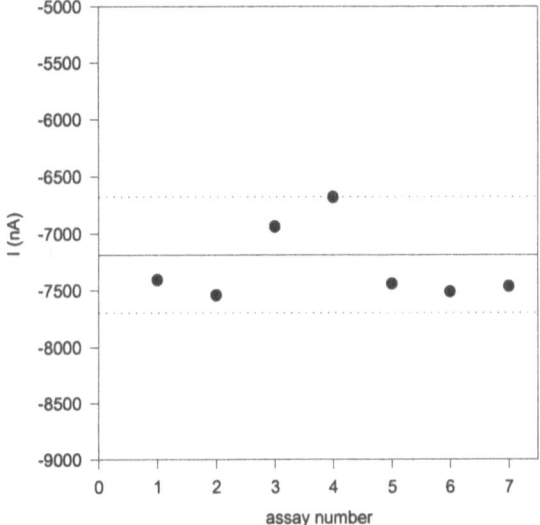

Figure 22: Reproducibility of the response of the immunosensor based on a RIgG-graphite-methacrylate composite containing 0.9 % (w/w) RIgG after successive incubations with a solution of GaRIgG peroxidase conjugate 0.63 % (v/v) in blocking buffer. The surface of the immunosensor was polished between assays. The solid line represents the mean value of the measurements. The broken lines mark the 99.5 % confidence interval. Intensity values were obtained after adding $4.7 \cdot 10^{-4}$ M H_2O_2

1.5.3   Magnetoimmunosensor

One of the strategies followed in immunoassay automation is to use flow injection analysis (FIA). The advantages of the FIA technique are the small sample volume needed and that analyses can be performed continuously without major instrumental complications. Therefore, automation may be implemented simply and economically compared to other systems using micropipettes, plates, etc. (Shellum and Gubitz 1989, Shellum and Gubitz 1993, Shulze et al. 1994, Locascio-Brown et al. 1990, Puchades et al. 1992).

FIA applications to immunoassay are based mainly on immunoreactors packed with solid particles, CPGs, gels, etc. that contain immobilized immunological material (Lee and Meyerhoff 198,; Shellum and Gubitz 1989, Locascio-Brown et al. 1990,

Stöcklein and Schmid 1990, Ivnitskii et al. 1992, Alwis and Wilson 1985, Alwis and Wilson 1987, Kelly and Christian 1982).

An important innovation is presented in the present work. An immunosensor is used instead of an immunoreactor separated from the sensing device. The immunological phase is located on the transducer. This arrangement means a greater integration of the experimental setup and a better sensitivity since the product of the enzyme reaction is not dispersed by the flow system but detected directly.

However, one of the main limitations for the use of immunosensors in flow systems is the difficulty of regenerating the solid phase. The high association constants call for the use of drastic conditions (high ionic force, extreme pH values) to break the immunological bond for a new immunoassay to take place. Furthermore, the immunological material has to behave reproducibly after being subjected to these extreme measures.

Table 10: Main features of a flow-injection magnetoimmunosensor system based on magnetic immuno-particles

VERSATILITY AND EASY MANIPULATION
Modified magnetic particles are easily manipulated, transported and stored, and allow their entrapment or release by means of a magnet.

REGENERATION
The surface of an immunosensor formed by magnetic particles modified with the immunological material is renewed for each analytical cycle by discarding the used particles.

REPRODUCIBILITY AND CONTROL OF THE ACTIVE SURFACE
The active surface of the immunosensor is reproducible and can be controlled by the volume of the injected magnetic particles.

EASY IMMOBILIZATION
The biological material is easily immobilized on the preactivated magnetic particles which allow the covalent bonding to proteins. There are commercial magnetic particles activated with different functional groups

EASILY AUTOMATED
Modified magnetic particles are easily integrated in a flow system and allow the automatization of immunoassay techniques.

To solve these problems, an immunosensor based on a renewable magnetic support has been developed in our laboratories. The immunoreagent is fixed to a field-effect transistor integrated to a flow injection system. The immunoreagent is immobilized on magnetic particles that contain a superparamagnetic material (γFe_2O_3 and Fe_3O_4). The surface of these particles can be modified to facilitate the immobilization of biological material. The particles are fixed or removed from the sensing area using a magnetic field. Other advantages of this material are shown in Table 10.

Using magnetic particles solves the problem of the regeneration of the immunological material which present other solid phases. This problem translates into a loss of sensitivity due to the use of chemical regeneration measures (Miller 1981). With magnetic particles it is possible to renovate the immunological material simply, economically and reproducibly.

The flow system designed and reported here can be modified to accommodate any type of detection, which is defined by the labeling agent used. The sensitivity of the method will depend on the type of label and the selected detection.

A potentiometric detection and the enzyme urease were chosen for the present work.

1.5.3.1 Conjugation of the magnetic particles. Construction of the immunosensor

The magnetic particles were obtained commercially activated with a tosyl group. This permits a covalent bond to be established with any type of protein by the following nucleophilic reaction.

80

The covalent bond ensures a good activity, stability and reproducibility. However, this type of bond can alter or block the active centers of the attached proteins.

The protocol developed for the present work ensures the immobilization of enzymes and antibodies to the magnetic particles without a loss of activity.

The magnetic particles (20 mg/ml) were incubated in a 0.1 M phosphate buffer solution at a pH 7.5 with 80 mg/ml rabbit immunoglobulin (RIgG). This incubation took place during 24 hours at 37 °C under gentle stirring. After this time, non-reacted groups were blocked by several wash cycles with BSA/PBS.

The magnetic particles conjugated with RIgG are injected to the flow system (Figure 23). They are retained with a neodymium magnet on the sensitive gate of a pH ISFET (Ion-Selective Field Effective Transistor) which forms the detection system. The magnetic particles immobilized on the pH ISFET form the immunosensor. The cell where the sensor is placed has a volume of 25 µl.

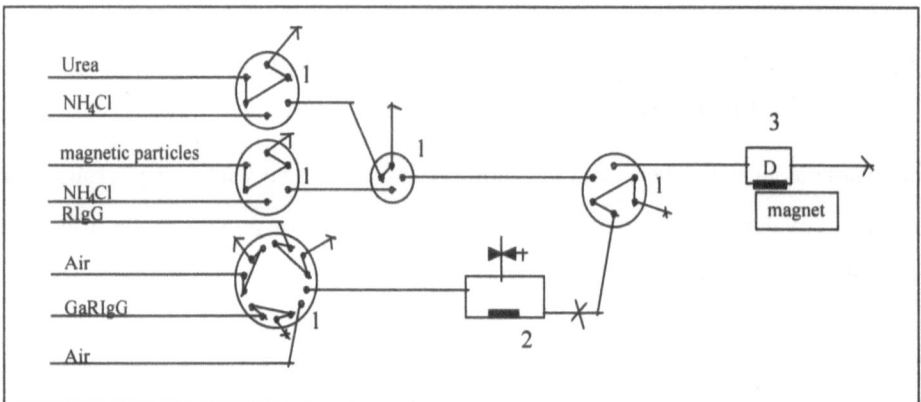

Figure 23: Schematic representation of the flow injection immunoassay (FIIA) system. (1) Injection valves; (2) preincubation chamber with magnetic stirring; (3) magnetoimmunosensor; (X) channel closed while the preincubation is done in the chamber

1.5.3.2   Label: Urease

The products obtained from the enzymatic catalysis of urease bring about a change of pH in the carrier solution. This pH change is detected by the ISFET according to the parameters shown in Table 9.

The carrier solution used was 0.1 M NH_4Cl. The main feature of this carrier is that the ammonia generated by the enzyme reaction transforms a segment of the carrier into a buffer system where pH varies according to: $pH = pK_a + \log ([NH_3]/[NH_4^+])$ (Alegret et al. 1990). If $[NH_4^+]$ is kept constant, any pH variation will be a linear function of $\log[NH_3]$ produced by the enzyme reaction and of the logarithm of the concentration of the substrate and therefore, of the concentration of the analyte.

1.5.3.3   Flow system

The developed flow system is shown in Figure 23. It accommodates all the stages of a competitive immunoassay. The magnetic particles modified with RIgG are injected on a NH_4Cl carrier solution. The particles are retained by a magnet once they reach the detector. The solutions of GaRIgG-urease and RIgG (analyte) are injected through a parallel channel on an air carrier. This carrier transports these solutions to a reaction chamber where they are preincubated for ten minutes using magnetic stirring. After this preincubation period has elapsed, the resulting solution is used to fill a 200 µl injection loop using a second pump. This solution is injected on the magnetic particles at a flow rate of 0.02 ml/min. This flow rate is sufficiently low to permit a second incubation with the RIgG immobilized on the magnetic particles.

In this second incubation, the GaRIgG-urease not bound to the analyte during the first incubation will become attached with the RIgG immobilized on the magnetic particles. After the magnetic particles have been discarded, no sizable peak is produced when different urea concentrations are injected. This is proof that the

Table 11: Experimental conditions for the determination of RIgG (competitive assay) using a magnetoimmunosensor integrated to a flow injection system. The electrochemical measurements were performed under the experimental conditions described in Table 9

Immunoconjugate label	UREASE	
Assay	COMPETITIVE	
Immunosensor preparation	Magnetic particles:	50μl magnetic particles (DYNABEADS M-280) 1.5 mg/ml
	Carrier solution flow rate:	0.2 ml/min
	Magnet:	activated
Preincubation	RIgG solution:	200 μl of several RIgG concentrations diluted from a stock solution with Tris 0.1 M, EDTA 0.001M i BSA 0.1 %
	GaRIgG urease solution:	200 μl of a GaRIgG-urease solution (1.2 % (v/v)) in Tris 0.1 M, EDTA 0.001 M at pH 7.5 with BSA 0.6 %
	Time:	10 minutes (77 % of the full signal is reached after 10 minutes)
Incubation	Volume of preincubated solution:	200 μl
	Carrier solution flow rate:	0.02 ml /min
	Carrier solution:	NH_4Cl 0.1 M
	Magnet:	activated
Detection	Enzymatic substrate solution:	Urea 0.1 M in NH_4Cl 0.1 M
	Injection volume:	50 μl
	Carrier solution flow rate:	0.2 ml/min
	Carrier solution:	NH_4Cl 0.1 M
	Magnet:	activated
Regeneration and rinsing	Carrier solution flow rate:	3 ml/min
	Carrier solution:	NH_4Cl 0.1 M
	Magnet:	deactivated

magnetic particles have not been retained by the system and no adsorption has occurred in the walls of the tubing or in any other parts of the system.

1.5.3.4   Competitive technique

A competitive immunoassay technique has been implemented using the flow system described above. The system measures RIgG with urease as a label enzyme using the conditions described in Table 11.

Figure 24 depicts the relative signal $(E/E_{max})\%$ *vs* $C_{analyte}$. A linear relationship between these variables is observed up to 0.332 mg/ml RIgG. The detection limit, defined as the concentration capable to produce a signal three times as large as the background noise, is 0.014 mg/ml.

This detection limit can be bettered choosing other labels in more sensitive detection setups such as those with amperometric transducers.

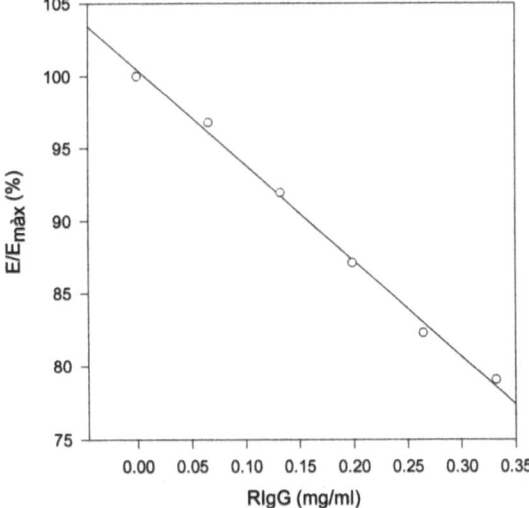

Figure 24: Calibration curve for the measurement of RIgG using a competitive technique and a flow system featuring a magnetoimmunosensor. Enzymatic substrate: urea 0.1 M. Further experimental details in Table 9 and 11. Regression: slope = -66 ± 6; y-intercept = 100 ± 1; r = 0.998

84

1.5.3.5 Renovation of the immunosensor surface

The advantage of using this type of immunosensors is the simple elimination of the solid phase. This is achieved by removing the restraining magnetic field. In this way a new and fresh surface is attained.

The reliability of the measurements depends on the reproducibility of the immunomagnetic phase. This can be evaluated by injecting successive batches of magnetic particles modified with RIgG to measure preincubated solutions containing RIgG (0.1992 mg/ml) and immunoconjugate following the procedure described above. The results show a good reproducibility with a relative standard deviation of 2 % (n=6) as shown in Figure 25.

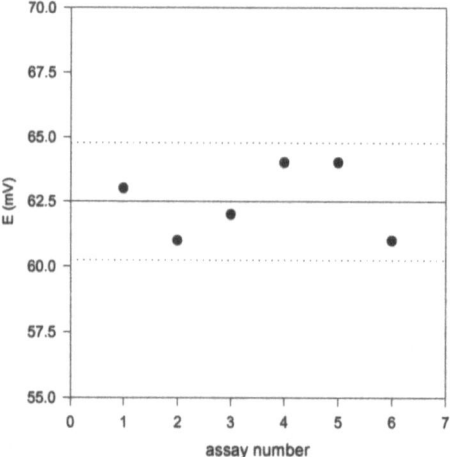

Figure 25: Reproducibility of the construction of magnetic immunosensors in a flow injection system. The microcell was filled with 50 µl of magnetic particles (1.5 mg/ml) for each assay. 200 µl of equal volumes of 1.2 % (v/v) GaRIgG urease conjugate and 0.20 mg/ml of RIgG were injected. Further experimental details are listed in Table 11. The solid line represents the mean value of the peak-heights measured. The broken lines mark the 99.5 % confidence interval. Enzymatic substrate: urea 0.1 M

1.5.4 Acknowledgements

This work has been funded by the European Commission, Directorate General for Science, Research and Development (XII-D-1), Environmental Technologies (EV5V-CT94-0407) and the Comisión Interministerial de Ciencia y Tecnología, Spain (BIO95-1196-CE and BIO96-0740).

1.5.5 References

Alegret S., Alonso J., Bartrolí J., del Valle M., Jaffrezic-Renault N., Duvault-Herrera, Y. (1990): Flow-through pH-ISFET as detector in the determination of ammonia. Anal. Chim. Acta 231, 53-58.

Alegret, S. (1996): Rigid carbon-polymer biocomposites for electrochemical sensing. A review. Analyst 121, 1751-1758.

Alegret, S., Alonso, J., Bartrolí, J., Céspedes, F., Martínez-Fàbregas, E., del Valle, M. (1996b): Amperometric biosensors based on bulk-modified epoxy graphite biocomposites. Sensors and Materials 8, 147-153.

Alegret, S., Céspedes, F., Martínez-Fàbregas E., Martorell, D., Morales, A., Centelles, D., Muñoz, J. (1996a): Carbon-polymer biocomposites for amperometric sensing. Biosensors & Bioelectronics 11, 35-44.

Alwis, W.U., Wilson, G.S. (1987): Rapid heterogeneous competitive electrochemical immunoassay for IgG in the picomole range . Anal. Chem. 59, 2786-2789.

Alwis, W.U., Wilson, G.S. (1985): Rapid sub-picomole electrochemical enzyme immunoassay for Immunoglobulin G. Anal. Chem. 57, 2754-2756.

Blake, C., Gould, B.J. (1984) : Use of enzymes in immunoassay techniques. A review. Analyst 109, 533-547.

Blanchard, G.C., Taylor, C.G., Busey, B.R., Williamson, J. (1990): Regeneration of immunosorbent surface used in clinical, industrial and environmental biosensors. J. Immunol. Methods 130, 263.

Buerk, D.G. (1993): Biosensors. Theory and applications. Technomic Publishing Company, USA.

Byfield, M.P., Abuknesha, R.A. (1994): Biochemical aspects of biosensors. Biosensors & Bioelectronics 9, 373-400.

Céspedes, F., Alegret, S. (1996): New materials for electrochemical sensing: glucose biosensors based on rigid carbon-polymer biocomposites. Food Technol. Biotechnol., Review 34, 143-146.

Céspedes, F., Martínez-Fàbregas, E., Alegret, S. (1993a): Amperometric glucose biosensor based on an electrcatalytically bulk-modified epoxy-graphite biocomposite. Anal. Chim. Acta 284, 21-26.

Céspedes, F., Martínez-Fàbregas, E., Bartrolí, J., Alegret, S. (1993b): Amperometric enzymatic glucose electrode based on an epoxy-graphite composite. Anal. Chim. Acta 273, 409-417.

Duan, Ch., Meyerhoff, M.E. (1994): Separation-free sandwich immunoassays using microporous gold electrodes and self-assembled monolayer/immobilized capture antibody. Anal Chem. 66, 1369-1377.

Dzantiev, A.V., Zherdev, A.V. (1996): Electrochemical immunosensors for determination of the pesticides 2,4-dichlorophenoxyacetic and 2,4,5-trichlorophenoxyacetic acids. Biosensors & Bioelectronics 11, 179-185.

Gascón, J., Martínez, E., Barceló, D. (1995): Determination of atrazine and alachlor in natural waters by a rapid-magnetic particle-based ELISA. Influence of common cross-reactants: deethylatrazine, deisopropylatrazine, simazine and metolachlor. Anal. Chim. Acta 311, 357-364.

Glazier, S.A., Rechnitz, G.A. (1991): Preparation and characterization of IgG-coated membranes for possible use as solid phases in enzyme immunosensors. Anal. Lett. 24, 1347-1362.

Gübitz, G., Shellum, C. (1993): Flow injection immunoassays. Anal. Chim. Acta 283, 421-428.

Hall, E.A.H. (1990): Biosensors. Open University Press, Biotechnology series, England.

Hammock, B.D., Mummas, R.O. (1980): Immunochemical technologies in environmental analysis, p. 321. In: Recent Advances in Pesticide Analytical Methodology (Harvey, J. R. L., Zweig, eds). ACS, Washington, DC.

Heinemann, W., Halsall, H.B. (1985): Strategies for electrochemical immunoassay. Anal. Chem. 57, 1321A-1331A.

Ivnitskii, D.M., Sitdikov, R.A., Kurochkin, V.E. (1992): Flow-injection amperometric system for enzyme immunoassay. Anal. Chim. Acta 261, 45-52.

Kaláb, T., Skládal, R. (1995): A disposable amperometric immunosensor for 2,4-dichlorophenoxyacetic acid. Anal. Chim. Acta 304, 361-368.

Kelly, T.A., Christian, G.D. (1982): Homogeneous enzymatic fluorescence immunoassay of serum IgG by continuous flow-injection analysis. Talanta 29, 1109-1112.

Killard, A.J., Deasy, B., O'Kennedy, R., Smyth, M.R. (1995): Antibodies. Production, functions and applications in biosensors. Trends in Anal. Chem. 14, 257-265.

Kindervater, R., Künnecke, W., Schmid, R.D. (1990): Exchangeable immobilized enzyme reactor for enzyme inhibition tests in flow-injection analysis using a magnetic device. Determination of pesticides in drinking water. Anal. Chim. Acta 234, 113-117.

Lee, I.H., Meyerhoff, M.E. (1988): Enzyme-linked flow-injection immunoassay using immobilized secondary antibody. Mikrochim. Acta III, 207-221.

Leech, D. (1994): Affinity biosensors. Chemical Society Reviews, 205-213.

Locascio-Brown, L., Plant, A.L., Horvath, V., Durst, R. (1990): Liposome flow-injection immunoassay: implications for sensitivity, dynamic range and antibody regeneration. Anal. Chem. 62, 2587-2593.

Marco, M.P., Gee, S., Hammock, B.D. (1995): Immunochemical techniques for environmental analysis. I. Immunosensors. Trends in Anal. Chem. 14, 341-350.

Martorell, D., Céspedes F., Martínez-Fàbregas, E., Alegret, S. (1994): Amperometric determination of pesticides using a biosensor based on a polishable graphite-epoxy biocomposite. Anal. Chim. Acta 290, 343-348.

Martorell, D., Céspedes, F., Martínez-Fàbregas, E., Alegret, S. (1997): Determination of organophosphorus and carbamate pesticides using a biosensor based on a polishable 7,7,8,8-tetracyanoquinodimethane-modified graphite-epoxy biocomposite. Anal. Chim. Acta 337, 305-313.

Miller J.R. (1981): Anal. Proc. 18, 264-267.

Morales, A., Céspedes, F., Muñoz, J., Martínez-Fàbregas, E., Alegret, S. (1996): Hydrogen peroxide amperometric biosensor based on a peroxidase-graphite-epoxy biocomposite. Anal. Chim. Acta 332, 131-138.

Nelson, J.O., Karu, A.E., Wong, R.B. (1995): Immunoanalysis of Agrochemicals, Emerging Technologies. ACS Symp. Series, Vol 586, ACS, Washington DC.

North, J.R. (1985): Immunosensors: antibody-based biosensors. Trends in Biotechnology 3, 180-186.

Owen, V.M. (1994): Market requirements for advanced biosensors in healthcare. Biosensors & Bioelectronics 9 (6), xxix-xxxiii.

Puchades, R., Maquieira, A., Atienza, J., Montoya, A. (1992): A comprehensive overview on the application of flow injection techniques in immunoanalysis. Crit. Rev. Analyt. Chem. 23, 301-321.

Rogers, K.R., Cao, C.J., Valdes, J.J., Eldfrawi, A.T., Eldfrawi, M.E. (1991): Fundam. Appl. Toxicol. 16, 810.

Rook, G.A.W., Cameron, C.H. (1981): J. Immunol. Methods 40, 109.

Sansubrino, A., Mascini, M. (1994): Development of an optical fibre sensor for ammonia, urea, urease and IgG. Biosensors & Bioelectronics 9, 207-216.

Santandreu, M., Céspedes, F., Alegret, S., Martínez-Fàbregas, E. (1997): Amperometric immunosensors based on rigid conducting immunocomposites. Anal. Chem. 69, 2080-2085.

Shellum, C., Gübitz, G. (1989): Flow-injection immunoassays with acridinium ester-based chemiluminescence detection. Anal. Chim. Acta 227, 97-107.

Shulze, B., Schlösser, A., Middendorf, C., Schelp, C., Schelper, T., Schügerl, K., Noé, W., Hoffmann, H., Howald, M. (1994): New immunoanalysis systems for application in biotechnology, p 717. In: Proceedings of the 6th European Congress on Biotechn. (Alberghina, L., Frontali, L., Sensi, P., eds). Elsevier Science.

Stöcklein, W., Schmid, R.D. (1990): Flow-injection immunoanalysis for the on-line monitoring of monoclonal antibodies. Anal. Chem. 234, 83-88.

Wang, J. (1988): Electroanalytical technology in clinical chemistry and laboratory medicine. VCH Publishers, New York. Chapter 1.

Wijesuriya, D., Breslin, K., Anderson, G., Shriver-Lake, L., Ligler, F.S. (1994): Regeneration of immobilized antibodies on fiber optic probes. Biosensors & Bioelectronics 9, 585-592.

2 Industrial Pollutants

2.1 Catalytic and Affinity Amperometric Biosensors for Phenols, Phosphates, and Atrazine: How Transduction Can Improve Performance

Arántzazu Narváez, Miguel Angel López, Elena González, Elena Domínguez, Juan José Fernández[*] and Ioanis Katakis[*]

Departamento de Química Analítica, Facultad de Farmacia, Universidad de Alcalá, 28871 Alcalá de Henares, Madrid, Spain

[*]Departament d'Enginyeria Química, Escola Tècnica Superior d'Enginyeria Química, Universitat Rovira i Virgili, 43006 Tarragona, Catalonia, Spain

Abstract. Three cases are presented where the rational design of transduction chemistries has led to improved catalytic and affinity electrochemical biosensors for environmental applications. Firstly, the improvement of the reductive recycling of tyrosinase-produced quinones by means of rational modification of electrode surfaces is demonstrated resulting in two orders of magnitude lowering of detection limits, and more than one order of magnitude improvement of the life time of phenolics sensors. Secondly, a phosphorylase A-phosphoglucomutase-glucose 6-phosphate dehydrogenase biosensor is demonstrated, that based on the use of this three-enzyme cascade and combined with new NADH oxidation mediators makes possible reagentless biosensors for phosphate detection. Thirdly, an immunosensor for atrazine is presented that based on electrochemically "wired" peroxidase-labelled atrazine and its competition for the binding sites of immobilised antibodies, reached $\mu g\ l^{-1}$ (ppb) detection limits and incubation times of minutes.

2.1.1 Introduction

Biosensors can have a significant impact on environmental monitoring especially when they are developed to replace analytical procedures that are technically difficult or to solve analytical problems inaccesible by traditional means. They can be especially attractive when they are developed as field instruments, probably not providing all the information necessary for an official method, but still enough to be used for routine testing and screening of samples by regulatory authorities and by industry. Such biosensors can be generic, giving an indication of the presence of some contaminant that has some toxic effects (for example toxicity sensors) or of a group of compounds (for example pesticides as inhibitors to the enzyme cholinesterase) or specific, taking advantage of the specificity of a biomolecule for some compound.

Our group has been involved in the development of biosensors for environmental monitoring for specific compounds or specific classes of compounds, using electrochemical transduction. The use of electrochemical biosensors has advantages especially from the point of view of application of the technology since electrochemical detection is low cost and technologically mature for field applications, since hand-held instruments are used widely in clinical diagnostics. Alternatively, electrochemical biosensors are easily incorporated in FIA systems at low capital investment, where they can be used as screening devices with high sample throughput. As new analytical tools, such biosensors must be reagentless, have appropriate detection limits, and show an acceptable reliability (reproducibility, accuracy, and stability, both shelve life and operational). Such characteristics can be improved by the rational manipulation of the transduction chemistries developed to achieve bio-electrocatalysis.

Tyrosinase, an enzyme with a relatively wide selectivity for phenolic compounds, is a good candidate for a biosensor correlating results with the aminoantipyrine detection method used by regulatory authorities as a phenol index. The efforts to use tyrosinase as an analytical tool for the detection of phenols and catechols include electrochemical transduction starting in the 1970's. The electroreduction of quinoid intermediates of the enzymatic reaction provides the additional advantages of enzyme activation and of the electroenzymatic recycling of catechol, a second enzymatic substrate and enzyme activator, giving rise to an amplification of the signal, and for this, still is the scheme of choice for many efforts in the literature to produce phenol sensors in aqueous or organic solvents. Still, the electroreduction step can be made more efficient on catalytically active surfaces, a fact that could be possible if a suitable mediator were incorporated on the electrode surface. The most successful surface modifications or mediators for such a purpose have been used by Kulys and Schmid (1990) who used tetracyanoquinodimethane (TCNQ) and more recently by Gründig et al. (1992), who used methylphenazonium which was incorporated by the same authors (Kotte et al. 1995) in a screen printed configuration based on mediator-activated zeolite included in the carbon-based ink. We have recently proposed (Hedenmo et al. 1997) the use of a phendione-based osmium mediator as surface modification chemistry for the improvement of the described electrocatalysis and recycling demonstrating a marked improvement of the electroanalytical performance of carbon paste tyrosinase electrodes, and here we demonstrate that such performance can be further improved by incorporating the surface modification to more catalytic electrode surfaces, and we report on the first efforts to evaluate these biosensors in real sample analysis.

A second analyte of relative interest for environmental monitoring is inorganic phosphate since it can be used as a measure of eutrophism, and the traditional method for its selective detection is cumbersome and requires chromatography, volumetric titrations or spectrophotometric determination. Therefore, the development of enzyme-based biosensors may be a highly selective and sensitive alternative to determine phosphate in environmental samples, fast and at low cost. Although various enzymatic biosensor configurations for phosphate have appeared in the literature in recent years (Guilbault & Nanjo 1975, Watanabe et al. 1988, Male & Luong 1991, Kulys et al. 1992, Wollenberger et al. 1992, Su & Mascini 1995) none of them is reagentless. Therefore, we have chosen a configuration first proposed by Guilbault and Cserfavi, (1976) based on the sequential action of three enzymes, phosphorylase A, phosphoglucomutase, and glucose 6-phosphate dehydrogenase, that due to the polymeric nature of the cosubstrate (glycogen) and the introduction of an osmium-based mediator catalysing the reversible oxidation of NADH (Hedenmo et al. 1996), opens the way to the construction of reagentless enzymatic phosphate sensors, and we report on the first results for the optimisation of this configuration.

Despite the convenience of automated ELISA approaches, amperometric immunosensors are still of interest because with the combination of microelectronics and electrode arrays they can be used as an alternative to spectrophotometric monitoring methods. Most of recently developed amperometric electrochemical immunosensors are based on the labelling of analytes with oxidoreductases and the competition between the labelled analyte and that in the sample for a limited number of immobilised antibodies. Mostly diffusional mediators have been used to shuttle electrons between the tracer enzyme and the electrode (Lu et al. 1996, Ivnitski and Rishpon 1996). The main problem associated with these mediators is leaching into the bulk solution. To improve sensitivity, we have used redox polymers with non-diffusional mediators that can transfer electrons from the redox centres of the tracer enzyme to the electrode. Vreeke et al. (1995) have reported the use of this so called "molecular wiring" for the detection of the avidin/biotin-HRP affinity reaction on the electrode surface. Furthermore, Lu et al. (1997) have reported an amperometric immunosensor for HRP involving an osmium redox polymer co-immobilised with anti-HRP antibody. We demonstrate here the use of a competitive heterogeneous amperometric affinity assay for the determination of the herbicide atrazine. The electrochemical immunosensor contains an Os-polymer that can efficiently transduce to electrons the peroxide flux through an antigen-peroxidase conjugate captured on the electrode surface by the immobilised antibody.

2.1.2 Materials and methods

2.1.2.1 Tyrosinase electrodes

Preparation of mediated carbon paste tyrosinase sensors has been described by Hedenmo et al. (1997). Surface modified tyrosinase electrodes were made using 3.1 mm spectrographic graphite rods (RW 001, Ringsdorff-Werke GmbH) preheated at 700 °C for 90 s and polished on wet, fine emery paper. Deposition of the osmium (4,4'-dimethyl 2,2'-bipyridine)$_2$(1,10-phenanthroline-5,6-dione) mediator (Osphendione) was made electrochemically by 50 consecutive scans between -500 and +700 mV. Immobilisation of tyrosinase was made via carbodiimide as described by Ortega et al. (1993) with incorporation of an equivalent amount of poly(vinyl pyridine) previously derivatized with bromoethylamine. The Osphendione mediator was synthesised as reported by Hedenmo et al. (1997). For Unmediated Electrodes (UEs) the same procedure was used omitting the electrochemical deposition of the mediator. Blank electrodes were similarly prepared but without immobilisation of the enzyme. All electrodes were stored in 0.1 M phosphate buffer pH 6.0 at 4 °C until use.

Measurements were made in a single-channel flow injection system with a wall-jet flow through amperometric cell connected to a three-electrode potentiostat with a saturated calomel reference electrode (SCE) and a Pt wire counter electrode. The mediated tyrosinase sensors were press-fitted into a Teflon holder and inserted as working electrode in the cell. The various parts of the flow injection system were connected with stainless steel tubing. Samples of 20 μl were injected into the carrier stream consisting of 0.1 M phosphate buffer pH 6.0. The flow rate was kept at 0.6 ml min^{-1}. All measurements were made at -90 mV vs. SCE. All samples were injected directly into the FIA system with no pretreatment at all. If necessary, dilution of samples was made with 0.1 M phosphate buffer pH 6.0.

Determination of phenol index (PI) was made by an external environmental analysis laboratory (Grupo Interlab, Madrid, Spain) with the 4-aminoantipyrine standard method (EPA 420.1 for water samples and modified 420.1 for soil samples). Soil samples were distilled (samples S1, S2, and S3) and water samples were extracted with chloroform (samples W1, W2, and W3).

2.1.2.2 Phosphate biosensors

Although potentially reagentless with immobilisation in a carbon paste or in a gel deposited on the electrode surface, all the preliminary optimisation results are reported with dialysis membrane-covered electrodes. Dialysis membranes (MWCO 12-14 000 D) from Medicell International Ltd. were pre-treated by heating in 10 mM EDTA solution and stored until use in 0.5% sodium azide solution. Dialysis membrane-covered electrodes were made by securing with an O-ring the membrane on the glassy carbon electrode surface (0.071 cm^2) on which had been previously deposited the appropriate amounts of enzymes, cofactors and mediators. Previous to the immobilisation step, always the electrode surface was polished with alumina (Buehler) of 20, 5 and 1 μm, sonicated, and checked for the absence of adsorbed electroactive species by cyclic voltammetry. Amperometric responses were measured with a three-electrode conventional electrochemical cell consisting of the working, auxiliary (stainless steel wire) and reference (SCE) electrodes, in a well stirred solution. The electrolyte was 0.05 M Tris Buffer, 0.2 M NaCl, containing the necessary cosubstrates unless mentioned otherwise. The response was controlled by a BAS Cyclic Voltammograph CV-1B potentiostat and the steady-state response was monitored with a stripchart recorder. Generally 6 electrodes were used for every data point and the standard deviation generally ranged from 8-25%, being lower for the electrodes containing more phosphoglucomutase. The $Os(1,10$-phenanthroline-5,6-dione$)_2Cl_2$ mediator used for the oxidation of NADH was synthesised by refluxing K_2OsCl_6 with the stoichiometric amount of the ligand in DMF.

2.1.2.3   Atrazine electrochemical immunosensors

The polyclonal anti-atrazine antibody was purchased from Chemicon. The atrazine standards were provided by Dr. Ehrenstorfer GmbH. A carboxylic triazine derivative, 4-chloro-6-(isopropylamino)-1,3,5-triazine-2-(4-aminobutyric acid) was synthesised and kindly provided by Dr. Pilar Marco (CSIC, Barcelona, Spain). Horseradish peroxidase type IV (HRP) and bovine serum albumin (BSA) were purchased from Sigma. Enzyme tracer was synthesised coupling the HRP enzyme to the atrazine derivative by the carbodiimide/n-hydroxysuccinimide procedure. The redox polymer PVP {Os (bpy)$_2$ Cl} was synthesised as previously described (Katakis and Heller, 1992). In addition, the following reagents were used: Poly (ethylene glycol 400 diglycidyl ether) (PEGDGE) (Polysciences). N,N-Dimethylformamide (Sharlau), N-Hydroxy-succinimide (Fluka), N,N'-Dicyclohexylcarbodiimide (DCC, Sigma), Polyoxyethylenesorbitan monolaurate (Tween 20, Sigma), H_2O_2 30% (Perhydrol, Merck). All other chemicals used were of analytical grade.

Hydrogen peroxide solutions were prepared daily. Incubation with standard atrazine and atrazine-HRP labelled was performed at room temperature in PBS (10 mM Na_2HPO_4, 1.5 mM KH_2PO_4, 2.7 mM KCl, 68 mM NaCl) containing 0.05% Tween 20, pH=7.4. Electrochemical measurements were performed at room temperature in phosphate buffered saline (PBS), pH = 7.4. Glassy carbon disk electrodes (3 mm diameter) were pre-treated as above. The electrochemical measurements were performed in a standard three-electrode cell with a platinum wire counter electrode and a Ag/AgCl Bioanalytical Systems reference electrode. The volume of the cell was 1 ml.

The best films were made by mixing 0.7 µl PVP-Os solution (5 g l^{-1}), 0.4 µl antibody commercial solution and 1 µl PEGDGE solution (2.5 g l^{-1}) and loaded onto a 3 mm vitreous carbon electrode. After drying under vacuum atmosphere, the films were cured for a minimum of 15 h at room temperature before use.

2.1.3   Results and discussion

The results indicate that in these three cases of biosensors for environmental applications, the rational design and use of transduction and immobilisation chemistries can lead to marked improvements in the analytical performance of the biosensors. More specifically, for each of the cases the most important results are summarised below.

2.1.3.1   Tyrosinase electrodes

We have reported (Hedenmo et al. 1997) that by the incorporation of the appropriate catalytic mediator as surface modifier in carbon paste electrodes it is possible to diminish the detection limits of such tyrosinase electrodes by two orders of magnitude, down to 10 nM (Å1 µg l^{-1}) and to increase their stability. After 5 hours of continuous phenol injections or during one week under storage conditions, the catalytic current of the mediated sensors was essentially unchanged, whereas the unmediated sensors lost 50% of their current in about 1 hour of operation. Presumably this improvement was achieved due to the change in the rate limiting step of the response from an electrochemical kinetic step to the mass transport limitation of oxygen or the phenolic substrate. However, when analysis is performed in drinking waters, the detection limits required by a phenolics sensor are lower. To further decrease the detection limit of the tyrosinase electrode we used a more catalytic surface, that of solid graphite electrodes immobilising the mediator by electrodeposition as described in the experimental section. The result was a detection limit of 1 nM as depicted in Figure 26 where the modified graphite and carbon paste electrode response is compared. The 1 nM detection limit is sufficient for use of the sensor in drinking water quality control.

With these improvements it was therefore thought that such sensors could be used for the detection of phenolic compounds in real environmental samples. The values obtained with the sensors represent the „biosensor index" (BI) meaning the overall amount of phenolic compounds detected in mg l^{-1}. Three soil samples and three water samples were analysed and the results compared with the phenol index in a blind test. The results are summarised in Table 12.

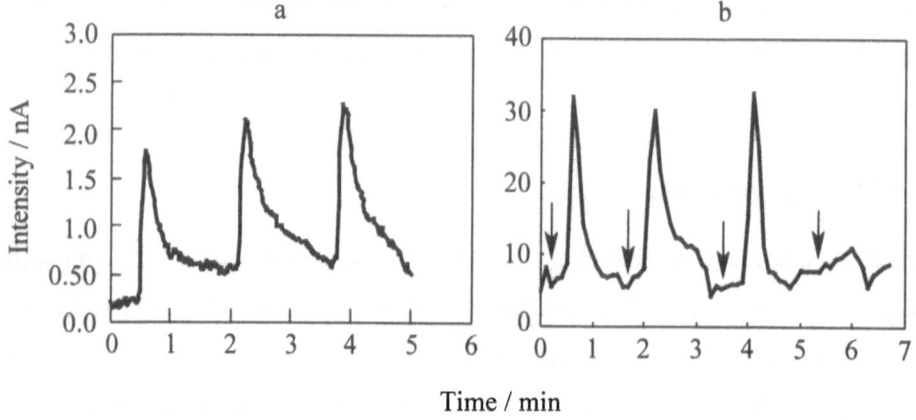

Figure 26: Response of (a) carbon paste and (b) graphite MEs to injections of 20 µl of 10 nM and 1 nM phenol respectively. Flow rate: 0.7 ml min^{-1} (a) and 0.6 ml min^{-1} (b), carrying buffer: 0.1 M phosphate buffer pH 6.0. Arrows denote injections. Fourth arrow in (b) is injection of blank

The following comments can be made for the "biosensor index" (BI) as compared to the phenol index (PI) of aminoantipyrine:

For the samples that were distilled (solid samples, S1, S2, S3) when the "biosensor index" was calculated by direct calibration of the sensor, a significant discrepancy with the PI values was observed particularly for S3. This discrepancy was attributed to the different response profile of the biosensor to that of the aminoantipyrine. Such differences were found for *p*-cresol, *m*-chlorophenol, catechol, *p*-chlorophenol, *m*- and *p*-hydroxybenzoic acids reaching differences of up to 6-8 times between the two methods. It was therefore decided that these samples should be analysed by the standard addition method, a fact that required five triplicate injections for every sample analysed. Even so, the time of analysis of each sample was less than one hour. The standard addition BI values agreed much better with the PI values. Presently, we have no explanation for the discrepancy with S1, but it might be attributed to some aminoantipyrine-interfering factor in the sample. For the water samples (W1, W2 and W3) when mediated electrodes were used, the interferences (readings of blank electrodes) were too high due to the high catalytic efficiency of the

Table 12: Comparison of the results obtained with the 4-aminoantipyrine method and the biosensor

Phenol Content / mg l^{-1}						
Method	S1	S2	S3	W1	W2	W3
4-AP	0.027	0.077	0.109	13.6	<0.005	0.038
Biosensor	0.077	0.088	0.049	11.6**	<0.0002**	0.050**
Biosensor*	0.044	0.074	0.114			

    *   these values were obtained with the standard addition method

    ** these values were obtained with unmediated electrodes

modified surface. It was therefore thought that UEs should be used in an effort to eliminate interferences. When this was tried, a very good agreement between BIs and PIs was observed even when direct calibration was used for the evaluation of the BI. It should be mentioned that most of these samples were coloured, a fact that made it necessary to extract the samples in chloroform for the PI determination. Such extraction was not necessary for the BI determination and all samples were injected directly. It should be noted that the <5 μg l^{-1} (ppb) evaluation of the aminoantipyrine method became <0.2 µg l^{-1} (ppb) for the biosensor (the detection limit). It has therefore been observed, that the real sample requirements dictated that unmediated electrodes be used that despite their higher limits of detection with laboratory samples, gave more reliable results with real samples. The uncontrolled amount of oxygen or the presence of interfering humic acids to both of which the mediated electrodes are more sensitive, are thought to be responsible for the poor results observed. To maintain the low detection limits while at the same time eliminating interferences, we are currently trying to incorporate a microdialysis membrane in the FIA system to eliminate some larger molecular weight interferents.

2.1.3.2 Phosphate biosensors

In the tri-enzymatic scheme used for phosphate detection Phosphorylase A (PA) produces glucose 1-phosphate from glycogen in the presence of phosphate. Phosphoglucomutase (PGM) produces glucose 6-phosphate that is oxidised by Glucose-6-phosphate Dehydrogenase (G6PDH) in the presence of NAD^+ producing

Glycogen + Pi $\xrightarrow{\text{PA}}$ Glucose-1-P

Glucose-1-P $\xrightarrow{\text{PGM}}$ Glucose-6-P

Glucose-6-P / 6-Phosphogluconate — G6PDH(ox) / G6PDH(red) — β-NAD(P)H / β-NAD(P)$^+$ — Med(ox) / Med(red) — e^- SCHEME I

NADH that is in turn oxidised by Os(1,10-phenanthroline-5,6-dione)$_2$Cl$_2$ that is recycled on the electrode surface at 200 mV vs. SCE as depicted in Scheme I.

The use of glycogen, a glucose polymer, allows for its easy inclusion in dialysis membrane-covered electrodes and promises the possibility of construction of reagentless phosphate electrodes. To this end we have examined the effect on the response of the other low-molecular weight cofactors, namely glucose 1,6 diphosphate and AMP, mentioned in the literature as necessary for enzymatic response. As shown in Figure 27, total exclusion of AMP could be achieved without any substantial loss of response, although the exclusion of glucose 1,6 diphosphate resulted in 70% loss of current due to its fundamental role in the equilibrium of phosphoglucomutase. To resolve this problem, we have changed the relative amounts of the enzymes included in the dialysis membrane and as is demonstrated in Table 13 it was possible to minimise the effect of glucose 1,6 diphosphate (G1,6-di-P) when the relative activities of the three enzymes were 0.5:1:0.4 for PA:PGM:G6PDH. A detection limit of 10 μM of phosphate was achieved with a linear dynamic range of up to 160 μM, and 2.0 μA cm^{-2} of maximum current density. The sensitivity in the linear region was of 3.3 μA cm^{-2} mM^{-1}.

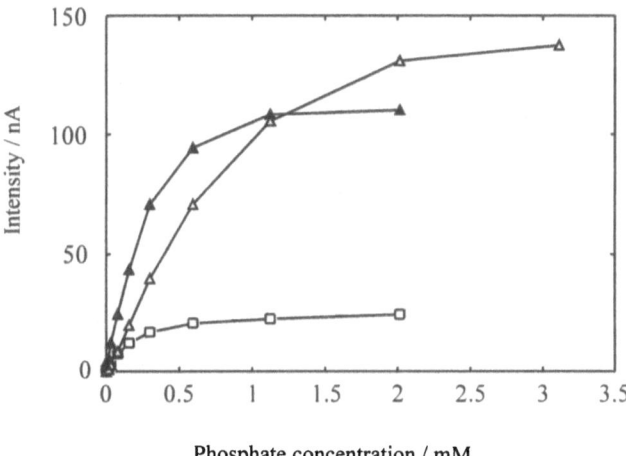

Phosphate concentration / mM

Figure 27: Effect of AMP and glucose 1,6 diphosphate on the response of phosphate biosensors containing in the dialysis membrane 0.48 units of phosphorylase A, 0.34 units of phosphoglucomutase, and 0.38 units of glucose 6-phosphate dehydrogenase, and 0.2 mg glycogen. ▲ in the presence of 0.1 M AMP and 23 μM glucose 1,6-diphosphate, △ in the absence of AMP and ❏ in the absence of glucose 1,6-diphosphate. Operating potential 0.2 V vs. SCE, 25°C, pH 7.0, vigorous stirring

Table 13:   Influence of relative enzyme activities on sensitivity and reagentless operation of phosphate biosensors

PA Units	PGM Units	G6PDH Units	Sensitivity $\mu A\ cm^{-2}\ mM^{-2}$	I_{max} nA	% of increase in response by G1,6-di-P
0.48	0.34	0.38	2.19	56	59
0.48	1.01	0.38	3.25	97	7
0.48	2.02	0.38	1.52	105	18
1.44	0.34	0.38	4.48	78	50
1.44	9.07	1.14	0.97	140	1

The electrodes presented a stable and maximum response at pH 6.5-7.5, had a response time of about 3 minutes and a half life time of 36 hours when stored at 4°C between tests. They showed no interference from NH_4^+, NO_3^-, NO_2^- and SO_4^{2-} although ClO_4^- and SCN^- did diminish the response. Currently we are examining the possibility to manufacture these electrodes in a carbon paste-type configuration without substantially altering their analytical characteristics.

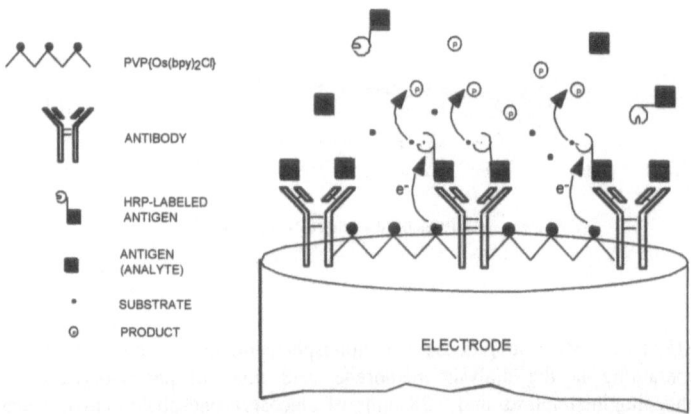

Figure 28: Transduction scheme of atrazine electrochemical immunoassay

2.1.3.3 Atrazine electrochemical immunosensors

The general transduction sequence and detection principle for these sensors are depicted in Figure 28.

Due to the polycationic character of the redox polymer it was expected to see a high degree of non-specific adsorption. However, it was detected that the non specific adsorption was less than 5% when the incubation buffer was not stirred, although slightly higher under stirring conditions (Parellada et al. 1998). The most important factor for the response of the electrodes was the efficiency of the electrochemistry as demonstrated in Figure 29a where the amount of "wire" included in the hydrogel is seen to significantly influence the response of the immunosensor. This increase in response with increased "wire" concentration in the hydrogel parallels the reversibility of the redox couple as observed by cyclic voltammetry (López et al. 1998).

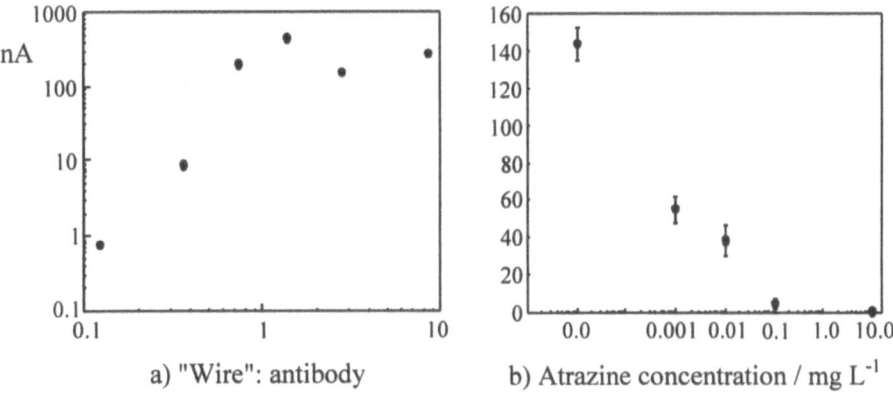

a) "Wire": antibody b) Atrazine concentration / mg L^{-1}

Figure 29: Characteristics of atrazine immunosensors. (a) Response of electrochemical immunosensor to HRP-labelled atrazine as a function of the "wire"/antibody ratio in the hydrogel. Incubation time was 30 minutes and response is monitored at 250 μM H_2O_2. (b) Standard calibration curve of electrochemical immunosensor to atrazine with 30 minutes incubation time in a mixture of HRP-labelled atrazine (a dilution of 1:50 of the conjugate, or approximately 10^{-5} M) and atrazine solution with the indicated concentration at "wire" : antibody ratio of 1.75

In Figure 29b a sample calibration curve is shown with the "best electrode" where it is shown that detection limits of the order of 1 ppb of atrazine concentration can easily be achieved. However, this limit of detection still remains three orders of magnitude higher than the achieved with standard ELISA techniques (Wittmann and Hock 1989) but for certain type of applications as for example when a portable or on-line device is needed, an immunosensor can be competitive with ELISA techniques. The results in Figure 29b were obtained with 30 min incubation of the electrochemical immunosensors, but it was verified that even with 5 minute trying to determine the reason for the low reproducibility between electrode responses with spectroelectro-chemical experiments.

2.1.4   Conclusions

It has been shown that the choice of transduction chemistries can improve the analytical characteristics of electrochemical biosensors based both on catalytic biosensing schemes (with reductive recycling of a reaction product or with direct cofactor oxidation) and affinity biosensors. Thus, chemically modified electrodes provide reversibility in the reductive recycling of quinone in tyrosinase electrodes resulting in low detection limit phenol sensors. However, additional considerations with respect to interfering substances dictated the most appropriate configuration for real sample analysis. NADH oxidising mediators can be used to drive sequential enzymatic reactions that yield potentially reagentless phosphate biosensors. The electrochemical transduction through molecular "wires" of a competition immunoassay is possible with short response times and $1 \ \mu g \ l^{-1}$ (ppb) limit of detection for an atrazine immunosensor.

2.1.5 Acknowledgements

Financial support from the Spanish CICYT (project ref: AMB 95-0075-C03), and the company SIDSA (ref. STQ 2408) through the SENIEC project of the GAME programme of the European Union are gratefully acknowledged. Grupo Interlab (Madrid, Spain) is also acknowledged for providing samples and standard analysis.

J.J.F. and A.N. acknowledge support from scholarships of the Universitat Rovira i Virgili (Department of Chemical Engineering) and the Spanish Ministry of Education and Culture (AP95 08986441), respectively.

2.1.6 References

Gründig, B., Strehlitz, B., Krabisch, C., Thielemann, H., Kotte, H., Gomoll, M., Kopinke, H., Pitzler, J. (1992): In: GBF-Monographs, Vol. 17. Biosensors: Fundamentals, Technologies and Applications (Schmid R.D.and Scheller F., eds). VCH, Weinheim. pp 275-285.

Guilbault, G., Nanjo, M. (1975): A phosphate-selective electrode based on immobilised alkaline phosphatase and glucose oxidase. Anal. Chim. Acta 78, 69-74.

Guilbault, G., Cserfalvi, T. (1976): Ion selective electrodes for phosphate using enzyme systems. Anal. Lett. 9, 277-289.

Hedenmo, M., Narváez, A., Domínguez, E., Katakis, I. (1996): Reagentless amperometric glucose dehydrogenase biosensor based on electrocatalytic oxidation of NADH by osmium phenanthroline dione mediator. The Analyst 121, 1891-1895.

Hedenmo, M., Narváez, A., Domínguez, E., Katakis, I. (1997): Improved mediated tyrosinase amperometric enzyme electrodes. J. Electroanal. Chem. 425, 1-11.

Ivnitski, D., Rishpon, J. (1996): A one-step, separation-free amperometric enzyme immunosensor. Biosens. Bioelectron. 11, 409-417.

Katakis, I., Heller, A. (1992): L-α-glycerophosphate and L-lactate electrodes based on the electrochemical "wiring" of oxidases. Anal. Chem. 64, 1008-1013.

Kotte, H., Gründig, B., Vorlop, K.-D., Strehlitz B., Stottmeister, U. (1995): Methylphenazonium-modified enzyme sensor based on polymer thick films for subnanomolar detection of phenols. Anal. Chem., 67, 65-70.

Kulys, J., Schmid, R.D. (1990): A sensitive enzyme electrode for phenol monitoring. Anal. Lett. 23, 589-597.

Kulys, J., Higgins, I., Bannister, J. (1992): Amperometric determination of phosphate ions by biosensor. Biosens. & Bioelectr. 7, 187-191.

López, M.A., Ortega, F., Domínguez, E., Katakis, I. (1998): Electrochemical immunosensor for the detection of atrazine. J. Mol. Recogn., submitted.

Lu, B., Iwuoha, E., Smyth, M., O'Kennedy, R. (1997): Development of an amperometric immunosensor for horseradish peroxidase (HRP) involving a non-diffusional osmium redox polymer co-immobilised with anti-HRP antibody. Anal. Commun. 34, 21-24.

Lu, B., Smyth, M., Quinn, J., Bogan, D., O'Kennedy, R. (1996): Development of a regenerable amperometric immunosensor for 7-Hydroxycoumarin. Electro-analysis 8, 619-622.

Male, K., Luong, H. (1991): An FIA biosensor system for the determination of phosphate. Biosens. & Bioelectr. 6, 581-587.

Ortega, F., Domínguez, E., Jönsson-Petterson G., Gorton, L. (1993): Amperometric determination of phenolic compounds using a tyrosinase graphite electrode in a flow system. J. Biotechnol. 31, 289-300.

Parellada, J., Narváez, A., López, M.A., Domínguez, E., Fernández, J.J., Pavlov, V., Katakis, I. (1998): Amperometric immunosensors and enzyme electrodes for environmental applications. Anal. Chim. Acta, submitted.

Su, Y., Mascini, M. (1995): AP-GOD biosensor based on a modified poly(phenol) film electrode and its application in the determination of low levels of phosphate. Anal. Lett. 28, 1359-1378.

Vreeke, M., Rocca, P., Heller, A. (1995): Direct electrical detection of dissolved biotinylated horseradish peroxidase, biotin, and avidin. Anal. Chem. 67, 303-306.

Watanabe, E., Endo, H., Toyama, K. (1988): Determination of phosphate ions with an enzyme sensor system. Biosens. & Bioelectr. 3, 297-306.

Wittmann, C., Hock, B. (1989): Improved enzyme immunoassay for the analysis of s-triazines in water samples. Food Agric. Immunol. 1, 211-224.

Wollenberger, U., Schubert, F., Scheller, F. (1992): Biosensor for sensitive phosphate detection. Sensors & Actuators B 7, 412-415.

2.2 A Master Analytical Protocol for Determining a Broad Spectrum of Organic Pollutants in Industrial Effluents

M. Castillo, A. Oubiña, J.S. Salau, J. Gascón and D. Barceló

Department of Environmental Chemistry, CID-CSIC, c/Jordi Girona 18-26, 08034 Barcelona, Spain

Abstract. Contaminated industrial effluents often contain a variety of organic pollutants which are difficult to analyze by standard GC/MS methods since often miss the more polar or nonvolatile of these organic compounds. The identification of highly polar analytes by chemical or rapid biological techniques is needed for characterization of the effluents. By correlating chemical characterization with toxicity a better picture of the effluent can be achieved.

The present work will point out a master analytical protocol for determining a broad spectrum of organic analytes present in various industrial effluents from Europe (petrochemical plant and hazardous waste). The protocol consisted in the setup of a methodology based on solid-phase extraction for the preconcentration of a variety of pollutants: catechol, nitrophenols, pentachlorophenol, benzidines, acridine, phosphates, phthalates and nonionic detergents (nonyl phenol) followed by LC-MS characterization using Atmospheric Pressure Chemical Ionization (APCI) in the Positive and Negative Ion modes.

The developed protocol permitted the unequivocal identification of target analytes like pentachlorophenol, tributyl phosphate, nonylphenol, dibutylphthalate, dimethylphthalate, bis(2-ethylhexyl)phthalate, tetramethyl thiourea, ethylbenzoate, 2-

methylbenzenesulfonamide, isothiocyanate-cyclohexane, neopentyl glycol and 1-methyl-2-pyrrolidinone present in industrial effluents.

Three RaPID-magnetic particle-based ELISA kits for determining pentachlorophenol, carcinogenic PAHs and BTEX (benzene, toluene, ethylbenzene and *o*-, *m*- and *p*-xylene) were also applied to the characterization of the industrial effluents .

2.2.1   Introduction

A great number of organic chemicals have been released to the environment. Many of them are toxic at low concentrations or may accumulate in sediments and organisms. As a consequence strict characterization of contaminated effluents needs to be done. In this respect the European Union (EU) has promulgated several years ago the so called "black list" of 132 dangerous substances (Directive 76/464/EC) of target analytes that should be monitored as dangerous substances discharged into the aquatic environment (Hennion et al. 1994). In the last year a new Directive on Integrated Pollution Prevention Control (IPPC) has been promulgated by the European Union (Directive 96/61/EC) expanding the range of pollutants that should be monitored in industrial effluents discharges like those from paper and pulp industry, refineries, textiles and many other sectors. In this perspective, research in the area of characterizing new pollutants in contaminated industrial effluents will be encouraged and will be expanded during the next coming years.

Common methods for identifying organic pollutants in contaminated industrial effluents involve generally the use of either dichloromethane liquid liquid extraction (LLE) or solid phase extraction (SPE) followed by gas chromatography-mass spectrometry (GC-MS) techniques with electron impact (EI) ionization (Burkhard et al. 1991, Benfenati et al. 1996, Betowski et al. 1996, Hale et al. 1996, Ellington et al. 1996) although few works have also been reported using chemical ionization

(Betowski et al. 1983, Barceló et al. 1990). By GC-MS a variety of non polar compounds are generally determined (Barceló et al. 1990, Burkhard et al. 1991, Benfenati et al. 1996, Hale et al. 1996, Ellington et al. 1996) but all the polar, ionic, heavy and thermally unstable compounds comprising more than 95% of the organic content (Betowski et al. 1996) cannot be analyzed. In this case, LC techniques are a good alternative as less sample clean up is required, thermally labile compounds are more easily analyzed, derivatization is usually not required and polar and high molecular weight compounds can be identified. However, the use of LC has been rarely reported in the characterization of polar analytes detected in industrial effluents. The US EPA has published two methods for the analysis of solid waste (SW-846), involving either particle beam (US EPA Method 8325) or thermospray (US EPA Method 8321). Nowadays the development of atmospheric pressure ionization (API) LC-MS interfaces allows to obtain similar structural information than chemical ionization techniques overcoming the limitations of other LC-MS interfacing devices like poor structural information or sensitivity such as for TSP and PB, respectively.

Environmental monitoring generally requires to analyze a large number of samples, therefore there is a need to search low cost, rapid and automated methods for analysis. In the last few years, one of the most developed field testing methods has been Enzyme Linked ImmunoSorbent Assays (ELISA) (Gascón et al. 1995, Oubiña et al. 1996) which have acquired a wide acceptance within the US especially. In this respect, the U.S. EPA has recently released the official method 4010A for the determination of pentachlorophenol (EPA method 4010A) among other methods for PCBs, PAHs, TNT and pesticides.

In the present work SPE with a polymeric sorbent followed by liquid chromatography and atmospheric pressure chemical ionization (APCI) mass spectrometry has been used for characterizing organic compounds in industrial effluents. Three RaPID-magnetic particle-based ELISA kits for determining pentachlorophenol, carcinogenic PAHs and BTEX (benzene, toluene, ethylbenzene and *o*-, *m*- and *p*-xylene) were also applied to the characterization of the industrial effluents. The final objective is to achieve a high level of knowledge about the composition and the concentration of pollutants present in contaminated industrial effluents in order to comply with the recently introduced directive.

2.2.2   Materials and methods

2.2.2.1   Chemicals and reagents

LC-grade solvents and pentachlorophenol, 2,4-dinitrophenol, 4-nitrophenol and 2-nitrophenol were obtained from Merck (Darmstadt, Germany). Catechol, phenol and *p*-cresol were from Sigma (St. Louis, MO, USA) and acridine, 2,2′-biphenol, 1-methylindol, benzophenone, ethylbenzoate, benzidines and phthalates were from Aldrich (Milwaukee, WI; USA). Tributylphosphate was obtained from Kodak (Rochester, NY, USA). 4-nonylphenol was from Kao Corporation (Barcelona, Spain). The rest of standards were a gift from Mario Negri Institute (Milano, Italy). Acetic acid pro-analysi grade from Panreac (Barcelona, Spain) was used. The RaPID-magnetic particle-based ELISA assays for pentachlorophenol, carcinogenic PAHs and BTEX from Ohmicron (Newtown, PA, USA) were purchased through MERCK (Barcelona, Spain). PBST was 0.2M phosphate buffer with 0.8% saline solution (pH 7.5) containing 0.05% Tween.

2.2.2.2   Sample collection

Samples from industrial effluents were collected in Pyrex borosilicate glass containers. Each bottle was rinsed with tap water and with high-purity water prior to sample addition. Sample preservation was accomplished by storing the bottles at 4 °C immediatly after sampling.

Treated plant effluent was collected at the discharge pipe (effluent) of a petrochemical plant during July '96. Two kinds of samples were supplied : A_1 from a new treatment plant in which chlorinated compounds were eliminated and A_2 from the conventional treatment plant.

Additional sampling was also conducted at an industrial landfille leachate, sample B and at a sugar refinery (sample C). More information about sample B was reported by Benfenati et al. (1996) using other analytical procedures.

2.2.2.3   Sample preparation

Off-line SPE experiments were performed using an Automated Sample Preparation with Extraction Columns system (ASPEC XL) fitted with an external 306 LC pump for the dispensing of samples through the SPE cartridges and with a 817 switching valve for the selection of samples from Gilson (Villiers-le-Bel, France). The drying step was carried out using a Baker Spe 12G apparatus from J. T. Baker (Deventer, Netherlands).

An off-line SPE method previously developed by our group (Castillo et al. in press) was used for preconcentration of sample A (A_1 and A_2). The styrene-divinylbenzene sorbent Lichrolut EN (200 mg, 6 ml) from Merck was used for off-line SPE purposes.

Sample B was toxicity based fractionated according to a protocol established by S. Galassi et al. (in preparation). 5 different extracts were supplied for analysis by LC-APCI-MS.

2.2.2.4   LC-APCI-MS conditions

For LC-APCI-MS experiments a VG Platform from Micromass (Manchester, UK) equipped with a standard atmospheric pressure ionization (API) source which can be configurated for APCI or ISP was used. The APCI interface consists of a heated nebulizer probe and a standard atmospheric pressure source equipped with a corona discharge needle. A detailed description of this system can be found elsewhere (Lacorte et al. 1996). The solvent was delivered by a Waters 616 gradient pump system controlled by a Waters 600 S controller from Waters-Millipore (US). Source and probe temperatures were set at 150 and 450 °C respectively, corona discharge voltage was maintained at 3 kV and the cone voltage was between 20 and 40 V. The HV lens voltage was set at 0.20 kV. In full scan mode the m/z range was from 80 to 400 in both negative ion (NI) mode and positive ion (PI) mode of ionization.

100 µl of the extracts were injected in the LC system using as mobile phase water and acetonitrile both acidified with 0.5% of acetic acid following the next gradient : from 30% of acetonitrile and 70% of water in isocratic conditions during 15 minutes to 100% of acetonitrile in 15 minutes and back to initial conditions in 5 min at a flow rate of 1 ml/min. An Hypersyl Green ENV column (150 mm x 4.6 mm i.d.) equipped with a guard column both from Shandon *HPLC* (Cheshire, UK) was used.

2.2.2.5   Immunoassay procedure

The samples were assayed according to the RaPID Assay package insert described by Oubiña et al. (in press) for pentachlorophenol determinations. The protocol for Carcinogenic PAHs and BTEX assays was very similar to the Pentachlorophenol assay, but the only difference was the first incubation time of 20 and 15 minutes, respectively, instead of 30 minutes. The spectrophotometric measurements were determined using the RPA-I RaPID Photometric Analyzer (Ohmicron, Newtown, PA, USA) whose detailed operations have been previously described by Rubio et al. (1991).

2.2.3   Results and discussion

2.2.3.1   Recoveries and breakthrough volumes

A preliminary study of SPE previously to the analysis of the samples was performed. In this sense the above mentioned off-line SPE method was applied for the preconcentration of ground water samples spiked with 50 µg/l of different phenolic compounds (catechol, phenol, 4-methylphenol, 2,4-dinitrophenol, 2,2′-biphenol, 4-nitrophenol, 2-nitrophenol, pentachlorophenol, 1-methylindol and naphthol), benzidines (3,3′-dichlorobenzidine, benzidine and 3,3′-dimethylbenzidine), acridine, benzophenone, phthalates (dibutylphthalate, dimethylphthalate and bis(2-ethylhexyl)phthalate), nonionic detergents (4-nonylphenol and ethoxylated 4-nonylphenol), tributylphosphate, ethylbenzoate, 1-methyl-2-pyrrolidinone, 1,1,3,3-tetramethyl-2-thiourea, isothiocyanate-cyclohexane, 2,2'-dimethyl-1,3-propanediol and 2-methylbenzenesulfonamide. Target compounds were chosen according to the origin of the samples and considering a compendium of contaminants commonly found in chemical disposal sites (Betowski et al. 1983, Benfenati et al. 1996, Betowski et al. 1996). Loading volumes of 300, 500 and 900 ml were preconcentrated (in triplicates) in the Lichrolut EN cartridges to evaluate the recoveries and breaktrough volumes of target compounds. Table 14 shows the main recoveries and relative standard deviation (RSD%) obtained for the target compounds in the preconcentration of different sample volumes using off-line SPE with Lichrolut EN followed by LC-UV with wavelength set at 280 and 310 nm.

Recoveries varying from 70 to 104% were obtained in the preconcentration of 900 ml for the most non-polar compounds (i.e. benzidines) although the most polar ones such as catechol, 2-methylbenzenesulfonamide and 2,2-dimethyl-1,3-propanediol were not detected due to breakthrough. Recoveries varying between 22 and 55% were obtained for these compounds in the preconcentration of 300 ml of waste water sample. Thus, results indicated that the detection of all compounds was only feasible for loading volumes of 300 ml at most.

In the present work, lower recoveries than those reported by Hagen et al. (1990) and Castillo et al. (in press) in tap and ground water, respectively were obtained. However, it should be noticed that industrial waste water containing interferences and particles that can decrease the effectiveness of the SPE process is used in the present work.

Table 14: Mean recoveries and RSD% (n=3) obtained in loading different volumes of industrial waste water spiked at 50 µg/l in the target analytes using off-line SPE followed by LC-UV at 280 and 310nm

Compound	V_{load}=300 ml	V_{load}=500 ml	V_{load}=900 ml
catechol	55 (13)	<5	<5
phenol	57 (17)	<5	<5
acridine	55 (15)	<5	<5
4-methylphenol	63 (11)	24 (11)	<5
2,4-dinitrophenol	59 (11)	32 (15)	23 (13)
2,2'-biphenol	88 (10)	66 (11)	43 (12)
4-nitrophenol	79 (13)	64 (12)	48 (13)
3,3'-dichlorobenzidine	108 (9)	104 (9)	51 (12)
2-nitrophenol	89 (11)	97 (12)	71 (13)
naphthol	97 (8)	107 (9)	69 (9)
benzidine	86 (9)	95 (8)	77 (9)
1-methylindol	92 (8)	97 (9)	95 (10)
benzophenone	105 (11)	104 (9)	103 (10)
3,3'-dimethylbenzidine	103 (9)	99 (10)	104 (10)
dibutylphthalate	54 (7)	48 (10)	41 (9)
dimethylphthalate	69 (15)	55 (13)	46 (9)
4-nonylphenol	34 (5)	26 (6)	13 (9)
pentachlorophenol	87 (9)	85 (8)	79 (11)
2-methylbenzenesulfonamide	22 (8)	9 (7)	<5
2,2-dimethyl-1,3-propanediol	24 (9)	11 (9)	<5
bis(2-ethylhexyl)phthalate	63 (9)	58 (10)	44 (9)

2.2.3.2 LC-APCI-MS

Characterization of water samples was carried out by means of LC-MS techniques. This approach allows to identify and to obtain information for the most polar, heavy and termally unstable pollutants which cannot be analyzed using GC techniques. An Atmospheric Pressure Chemical Ionization interface was used and its efficiency to detect target compounds was checked before analyzing waste water samples. Preliminary flow injection analysis (FIA) were performed in order to establish the analysis conditions. In this sense, two ionization modes (positive-ion and negative-ion) and different cone voltage (20 and 40V) were tested. In general, more fragmentation was obtained via collision induced dissociation (CID) by raising the cone voltage which allows to get structural information for the identification of unknowns. However, extraction voltage of 20V was preferred because it led to higher sensitivity than 40V with enough structural information in both NI and PI modes. It is interesting to note that $[M+H]^+$ and $[M-H]^-$ ions were detected in PI and NI modes respectively for almost all target compounds (except isothiocyanate-cyclohexane in PI mode and 4-nonylphenol in PI and NI mode). Thus indicating that this technique allows to obtain important molecular weight information.

The developed method of SPE followed by LC-APCI-MS was applied to the analysis of industrial effluents (A_1 and A_2) and samples from landfill B. Both NI and PI modes with extraction voltage set at 20 and 40V were used for real samples in order to detect as much compounds as possible. All compounds were identified by matching retention time and MS spectrum in PI and NI modes with authentic standards.

Total Organic Carbon was also measured in samples A_1 and A_2 with values of 13 and 25 mgC/l which are higher as compared to surface waters used for drinking purposes (4 - 7 mg C/l) but are much lower than than those found in highly contaminated industrial wastes (Betowski et al. 1996).

The chromatographic profile obtained in full scan mode showed chromatographic peaks containing unresolved components. Nevertheless, SIM chromatograms provided a more selective method avoiding the presence of interferences. The full-scan LC-MS chromatogram and SIM LC-MS chromatogram for m/z=263 under NI conditions at 40 V obtained by injecting 100 μl of the extract resulting from the preconcentration of 200 ml of sample A_2 in Lichrolut EN is shown in Figure 30. Unequivocal identification of pentachlorophenol (PCP) was possible.

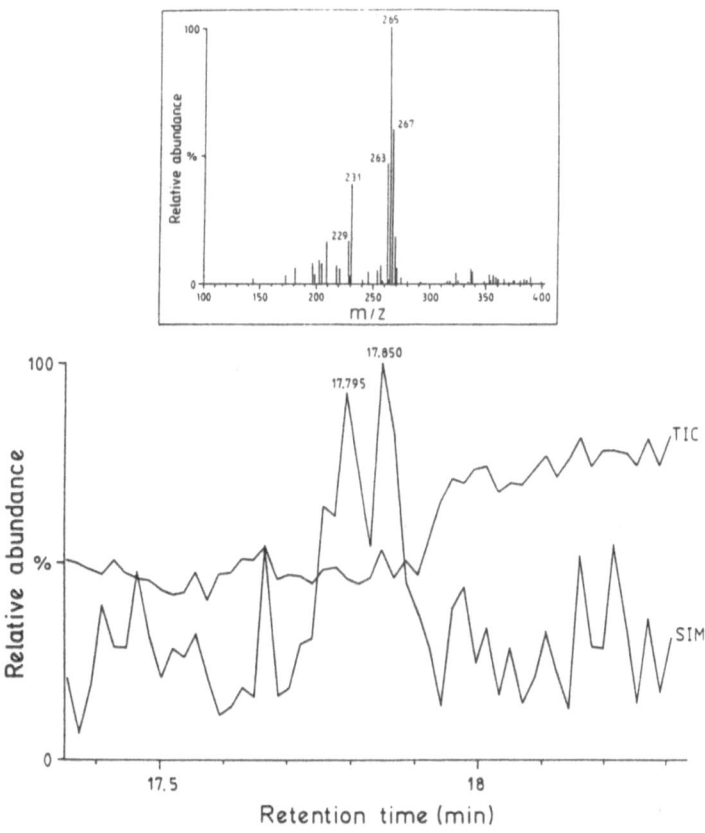

Figure 30: Full-scan LC-MS chromatogram and SIM LC-MS chromatogram for m/z=263 obtained with extraction voltage set at 20V under NI conditions in the injection of 100 μl of the extract of sample A_1. LC-APCI-MS spectrum for pentachlorophenol

PCP is frequently formed in bleaching processes and its presence in industrial effluents has been reported (Oubiña et al. in press, Sharma et al. 1996, Wilkins et al. 1996) although its degradation products such as less substituted chlorophenols are more frequently detected (Hale et al. 1996, Ellington et al. 1996, Oubiña et al. in press). Pentachlorophenol is on the Hazardous Substance List because it is regulated by OSHA (the legal airbone permissible exposure limit (PEL) is 0.5 mg/m^3 averaged over an 8 hour workshift) and cited by ACGIH. Its oral LD_{50}-for rats is 25-200 mg/Kg, thus this compound is classified as a toxic one (Wagner et al. 1983).

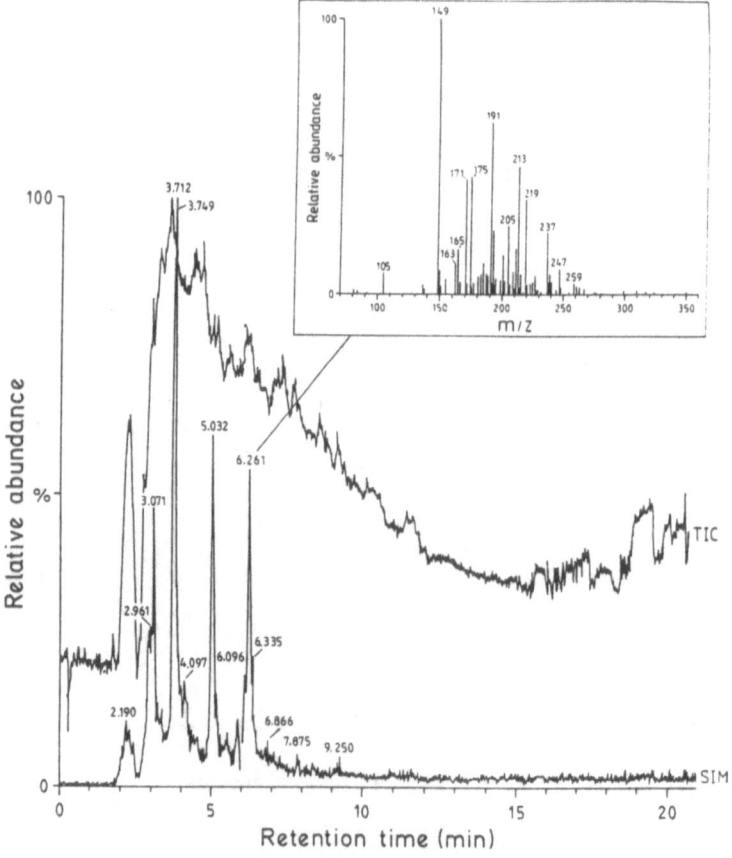

Figure 31: Full-scan LC-MS chromatogram and SIM LC-MS chromatogram for m/z=149 obtained with extraction voltage set at 20V under NI conditions in the injection of 100 μl of the extract of sample A$_1$. LC-APCI-MS spectrum for nonylphenol

LC-APCI-MS chromatogram in SIM mode for m/z=149 (see Figure 31) showed the presence of 4-nonylphenol in sample A_2 which was confirmed by injecting an authentic standard.

Table 15: Retrieved compounds and their estimated LODs and concentrations (results corrected for recovery) identified in industrial waste waters using LC-APCI-MS. n.d.: not determined

Compounds	Sample	LOD. (μg/l)	Conc. (μg/l)
Tributylphosphate	B	0.17	0.66
Dibutylphthalate	B	0.08	0.78
Dimethylphthalate	A_2	0.06	0.60
Ethylbenzoate	A_1	3.82	51.0
	B	n.d.	54.5
4-Nonylphenol	A_2	2.91	12.0
Pentachlorophenol	A_2	0.37	0.4
1-Methyl-2-pyrrolidinone	B	1.12	0.16
2-Methylbenzenesulfonamide	B	1.88	0.17
1,1,3,3-Tetramethyl-2-thiourea	A_2	1.97	39.3
Isothiocyanate-cyclohexane	A_1	0.96	11.2
2,2-Dimethyl-1,3-propanediol	B	1.00	14.7
Bis(2-ethylhexyl)phthalate	B	0.10	3.0

Nonylphenols are degradation products of nonylphenol ethoxylates (nonionic surfactants) often added to separators to breakup oil/water emulsions. They have also been used as plastic additives (Soto et al. 1991) or in Matacil (insecticide) formulations (Li et al. 1981) although these inputs are relatively minor. 4-nonylphenol is persistent, lipophilic and toxic to aquatic organisms with a 96-h LC_{50} values for salmon and trout ranging from 0.13 to 0.23 mg/l (Lee et al. 1995). It has been recently found to be estrogenic (Soto et al. 1991, Kirk-Othmer's Encyclopaedia of Chemical Technology).

Table 15 lists all the compounds observed in samples from sites A and B and their estimated concentrations and limits of detection (LOD) for the developed off-line SPE coupled to LC-APCI-MS method. The highest concentration level was obtained for ethylbenzoate, nevertheless this is not of concern due to the low toxicity of this contaminant and the absence of legislation. On the contrary, it stands out the presence of 1,1,3,3-tetramethyl-2-thiourea which is listed on the US EPA list of toxic substances. Thiourea compounds are mainly used as accelerators in the rubber industry and they are frequently found in PVC plastic or adhesive, diazo paper and paints or glue remover. The rest of pollutants are present in lower concentrations. Nevertheless, all of them (except ethylbenzoate) are included in the lists regulated by the US EPA, OSHA or EC and therefore it is necessary to control pollution in order to prevent an alarming level.

2.2.3.3   ELISA determinations

The analytical interference due to the limited selectivity of antibodies, commonly referred to as cross-reactivity (CR), is one of the most important aspects to take into account to correctly estimate the final values of the ELISA determinations (Miller et al. 1992, Oubiña et al. in press). In order to characterize the selectivity of the antibodies, a battery of cross-reactants (structurally similar compounds that may compete for the antibody binding sites) had to be checked. In our case, the selected analytes were obtained from the data reported in Table 15. The list include two phthalates (dibutylphthalate and dimethylphthalate), a nonionic detergent, 4-nonylphenol, tributylphosphate, ethylbenzoate, 1-methyl-2-pyrrolidinone, 1,1,3,3-tetramethyl-2-thiourea, isothiocyanate-cyclohexane, 2,2'-dimethyl-1,3-propanediol and 2-methylbenzenesulfonamide were tested in the Pentachlorophenol and

Carcinogenic PAHs tests, to be sure to detect the possible increase of the final concentration of the target analytes.

Table 16: Mean recoveries (in µg/l) and coefficients of variation (in percentage, n=3) obtained with water spiked at the same concentration found in the waste water samples (see Table 15) using the Pentachlorophenol and the Carcinogenic PAHs RaPID Assay

Compounds	% Recovery	
	Pentachlorophenol RaPID Assay	Carcinogenic PAHs RaPID Assay
Tributylphosphate	0.11 (5)	37 (2.6)
Dibutylphthalate	-	31 (6.2)
Dimethylphthalate	-	42 (10)
Ethylbenzoate	-	0.16 (8.9)
4-Nonylphenol	0.75 (2.9)	2 (11)
Pentachlorophenol	100 (1)	-
1-Methyl-2-pyrrolidinone	47 (10)	147 (16)
2-Methylbenzenesulfonamide	77 (12)	124 (5.8)
1,1,3,3-Tetramethyl-2-thiourea	0.3 (1.6)	0.68 (3.0)
Isothiocyanate-cyclohexane	-	1.45 (6.9)
2,2-Dimethyl-1,3-propanediol	-	0.93 (2.3)

Table 16 shows the mean responses using the two tests obtained with ground water spiked at the same concentration level found in the waste water samples of Table 15. The most remarkable data correspond to the Carcinogenic PAHs kit, because all the studied compounds could be detected, whereas the pentachlorophenol kit only responded to six of them (pentachlorophenol, ethylbenzoate, 4-nonylphenol, 1-methyl-2-pyrrolidinone, 2-methylbenzenesulfon-amide and 1,1,3,3-tetramethyl-2-thiourea).

Table 17 lists the values obtained in samples from sites A, B and C using the Pentachlorophenol, Carcinogenic PAHs and BTEX ELISA tests. The BTEX test detects petroleum hydrocarbons (benzene, toluene, ethylbenzene and m-, o- and *p*-xylene) expressed as the sum of the six components. The total BTEX RaPID Assay presents a sensitivity, Least Detectable Dose (LDD) based on a 90% B/B_0, of 0.02 mg/l. The carcinogenic PAHs RaPID Assay detects carcinogenic PAHs and related compounds to different degrees referred in the cross-reactivity data of the kit. So the possible cross-reactants can increase this final concentration. The sensitivity of this assay (as benzo [a] anthracene) in water is 0.01 µg/l. Pentachlorophenol ELISA test can also detect a variety of related compounds, especially chlorophenols as it was shown in a cross-reactivity study using this kit (Oubiña et al. in press).

Table 17: Concentrations of pentachlorophenol, carcinogenic PAHs and petroleum hydrocarbons (commonly referred to as BTEX-benzene, toluene, ethylbenzene and xylenes) found in four industrial waste waters using the Pentachlorophenol, Carcinogenic PAHs and BTEX RaPID ELISA. A: Waste water samples of a petrochemical plant during July '96. A_1: from a new treatment plant in which chlorinated compounds were eliminated. A_2: from the conventional treatment plant. B: Waste water sample from an industrial leachate. C: Waste water sample of a sugar refinery. [1]Values expressed in µg/l. [2]Values expressed in mg/l. [2]Total BTEX, equivalent parts of benzene, toluene, ethylbenzene and m-, o- and p-xylene expressed as the sum of the six components. [3]Values below detection limit. Coefficient of variation (in %, n = 6)

Samples	Pentachlorophenol[1]	Carcinogenic PAHs[1]	BTEX[2]
A_1	0.21 (3)	3.4 (11)	n.d.[3]
A_2	0.32 (3)	1.88 (8)	n.d.
B	4.43 (2)	7.8 (8)	15.3 (8)
C	6.91 (10)	18.4 (13)	7.14 (7)

The four waste water samples could be analyzed directly by immunoassay without previous filtration or clean-up. BTEX could only be detected in samples B and C whereas all samples gave a positive answer with the pentachlorophenol and carcinogenic PAHs kits. Considering the reported data in Table 16, the values of Table 17 correspond, not only to the target analytes, but also to the cross-reactants present in the waste water samples. So, in this way, ELISA determinations can be used as a measurement of the related contamination present in the waste water samples. In this respect, and taking into account the relatively high values of carcinogenic PAHs obtained by ELISA (see Table 17) and as no PAHs were detected in those samples (see Table 15), these values can be attributed in a first instance to phthalates that exhibit high cross-reactivity (see Table 16).

The good correlation between chromatographic and ELISA determinations could be noticed in sample A_2. Pentachlorophenol was detected in this sample by LC-APCI-MS at 0.4 µg/l, whereas 0.32 µg/l were obtained by ELISA. In this case, although the pentachlorophenol should be detected at 100%, the presence of other cross-reactants, 4-nonylphenol and 1,1,3,3-tetramethyl-2-thiourea can interfere the detection of the target analyte. As reported by Oubiña et al. (in press) the percentage of the cross-reactivity can vary versus the cross-reactant concentrations, showing low values at higher doses of these cross-reactants.

Another important point to take into account is the absence of previous treatments like filtration or clean-up, showing that these results can only be used as a screening tool of waste water samples and for giving a rapid estimation of the contamination of these waste water effluents.

2.2.4 Conclusions

The application of an LC-APCI-MS technique to the characterization of industrial waste waters and the use of immunoassay procedures for their screening are presented. LC-APCI-MS combined with SPE permits to achieve a good sensitivity of the method with only 200-300 ml of waste water samples. The incomplete characterization of these samples was linked to the lack of standards necessary to identify and quantify all the chromatographic peaks.

Three different kits, Pentachlorophenol, Carcinogenic PAHs and BTEX RaPID Assay have been used for screening purposes of four waste water samples. Even though overestimation could be observed due to the absence of treatments in the ELISA determinations and the interferences due to the cross-reactants detected by LC-APCI-MS, the combination of both techniques, automated systems like ASPEC XL for SPE followed by LC-MS and immunoassays, open a new window for monitoring organic compounds present in industrial effluents. Further work in this area should involve the issue of biosensors for a rapid determination of toxic analytes present in contaminated industrial effluents, which is one of the priorities of the present Environmental and Climate Program of the EU.

2.2.5   Acknowledgements

This work has been supported by the projects entitled : "Protocol for the Evaluation of Residues in Industrial Contaminated Liquid Effluents (Pericles)" [Contract ENV4-CT95-0021] from the Environment and Climate Program, 1994-98, Commission of the European Communities and CICYT (AMB96-1675-CE).
We thank Merck for supplying the SPE cartridges.
S. Galassi, E. Benfenati and F. Ventura are thanked for providing toxicologically fractionated industrial waste samples, standards and for carrying out the TOC measurements, respectively.

2.2.6   References

Amdur (1991): Cassaret and Doull's Toxicology, p. 499. In: The Basic Science of Poisons.

Barceló, D., Porte, C., Cid, J., Albaigés, J. (1990): Determination of organophosphorus compounds in mediterranean coastal waters and biota samples using gas chromatography with nitrogen-phosphorus and chemical ionization mass spectrometric detection. Intern. J. Environ. Anal. Chem. 38, 199-209.

Benfenati, E., Facchini, G., Pierucci, P., Fanelli, R. (1996): Identification of organic contaminants in leachates from industrial waste landfills. Trends Anal. Chem. 15, 305-310.

Betowski, L.D., Kendall, D.S., Pace, C.M., Donnelly, J.R. (1996): Characterization of groundwater samples from superfund sites by gas chromatography/mass spectrometry and liquid chromatography/mass spectrometry. Environ. Sci. Technol. 30, 3558-3564.

Betowski, L.D., Webb, H.M., Sauter, A.D. (1983): Pulsed positive ion negative ion chemical ionization mass spectrometric applications to environmental and hazardous waste analysis. Biom. Mass Spectrom. 10, 369- 376.

Burkhard, L.P., Durhan, E.J., Lukasewycz, M.T. (1991): Identification of nonpolar toxicants in effluents using toxicity-based fractionation with gas chromatography/ mass spectrometry. Anal. Chem. 63, 277-283.

Castillo, M., Puig, D., Barceló, D. (1997): Determination of priority phenolic compounds in water and industrial effluents by polymeric liquid-solid extraction cartridges using automated sample preparation with extraction columns and liquid chromatography. Use of liquid-solic extraction cartridges for stabilization of phenols. J. Chromatogr. A 778, 301-311.

Directive 96/61/EC, OJ No L 257 of 10.10.1996.

Directive 76/464/EC Ellington, J.J., Thurston, R.V., Sukyte, J., Kvietkus, K. (1996): Hazardous chemicals in waters of Lithuania. Trends Anal. Chem. 15, 215-224.

Galassi, S., Provini, A. in preparation.

Gascón, J., Durand, G., Barceló, D. (1995): Pilot survey for atrazine and total chlorotriazines in estuarine waters using magnetic particle-based immunoassay and gas chromatography-nitrogen/phosphorus detection. Environ. Sci. Technol. 29, 1551-1556.

Gee, S., Hammock, B.D., Van Emmon, J. (1994): A user's guide to environmental immunochemical analysis. U.S. Environmental Protection Agency: Las Vegas, Nev, EPA/540/r-94/509.

Hagen, D.F., Markell, C.G., Schmitt, G.A., Blevins, D.D. (1990): Membrane approach to solid-phase extractions. Anal. Chim. Acta 236, 160.

Hale, R.C., Smith, C.L. (1996): A multiresidue approach for trace organic pollutants: application to effluents and associated aquatic sediments and biota from the southern chesapeake bay drainage basin 1985-1992. Intern. J. Environ. Anal. Chem. 64, 21-33.

Hennion, M.-C., Pichon, V., Barceló, D. (1994): Surface water analysis (trace-organic contaminants) and EC regulations. Trends Anal. Chem. 13, 361-372.

Jobling, S., Reynolds, T., White, R., Parker, M.G., Sumpter, J.P. (1995): A variety of environmentally persistent chemicals, including some phthalate plasticizers, are weakly estrogenic. Environ. Health Persp. 103, 582-587.

Kirk-Othmer's Encyclopaedia of Chemical Technology, 3rd ed., Vol.18.

Lacorte, S., Barceló, D. (1996): Determination of parts per trillion levels of organophosphorus pesticides in groundwater by automated on-line liquid-solid extraction followed by liquid chormatography/atmospheric pressure chemical ionization mass spectrometry using positive and negative ion modes of operation. Anal. Chem. 68, 2466.

Lee, H.-B., Peart, T.E. (1995): Determination of 4-nonylphenol in effluent and sludge from sewage treatment plants. Anal. Chem. 67, 1976.

Li, H.F., Zwicker, R. (1981): Toxicity of Matacil, an insecticide, assessed on cultivated cells. In Vitro 17, 202.

MAFF 1995. Food Surveillance information sheet number 60: Phthalates in paper and board packaging. UK Ministry of Agriculture, Fisheries and Food.

Miller, J.J., Valdes, R., (1992): Methods for calculating cross-reactivity in immunoassays. Journal of Clinical Immunoassay. 15, 97-107.

Oubiña, A., Ferrer, I., Gascón, J., Barceló, D. (1996): Disappearance of aerially applied fenitrothion in rice crop waters. Environ. Sci. Technol. 30, 3551- 3557.

Oubiña, A., Gascón, J., Barceló, D. (1997): Multianalyte effect in the determination of cross-reactivities of pesticide immunoassays in water matrices. Anal. Chim. Acta 347, 121-130.

Oubiña, A., Gascón, J., Ferrer, I., Barceló, D. (1996): Evaluation of a magnetic particle-based ELISA for the determination of chlorpyrifos-ethyl in natural waters and soil samples. Environ. Sci. Technol. 30, 509-512.

Oubiña, A., Puig, D., Gascón, J., Barceló, D. (1997): Determination of pentachloro-phenol in certified waste waters, soils samples and industrial effluents using ELISA and liquid solid estraction followed by liquid chromatography. Anal. Chim. Acta. 346, 49-59.

Rubio, F.M., Itak, J.A., Scutellaro, A.M., Selisker, S.Y., Herzog, D.P. (1991): Performance characteristics for a novel magnetic-particle-based enzyme-linked immmunosorbent assay for the quantitative analysis of atrazine and related triazines in water samples. Food Agric. Immunol. 3, 113-125.

Schneider, P., Gee, S.J., Kreissig, S.B., Kraemer, P., Marco, M.P., Lucas, A.D., Hammock, B.D. (1995): Troubleshooting during the development and use of immunoassays for environmental analysis, p. 103-122. In: New Frontiers in Agrochemical Immunoassay (Kurtz, D.A., Skerrit, J.H., Stanker, L., eds). AOAC International, Arlington, Va, USA.

Sharma, C., Mohanty, S., Kumar, S., Rao, N.J. (1996): Gas chromatographic analysis of chlorophenolic, resin and fatty acids in effluents from bleaching processes of agricultural residues. Intern. J. Environ. Anal. Chem. 64, 289-300.

Soto, A.M., Justicia, H., Wray, J.W. (1991): p-Nonyl-phenol: and estrogenic xenobiotic released from „modified" polystyrene. Environ. Health Perspect. 92, 167-173.

US EPA Method 8321A. Solvent extractable non volatile compounds by high performance liquid chromatography/particle beam/mass spectrometry (HPLC/PB/MS) or ultraviolet (UV) detection.

US EPA, Office of Solid Waste and Emergency response, Washington DC, pp 1-50, (1995).

US EPA Method 8325. Solvent extractable non volatile compounds by high performance liquid chromatography/thermospray/mass spectrometry (HPLC/TS/MS) or ultraviolet (UV) detection. US EPA, Office of Solid Waste and Emergency response, Washington DC, pp 1-50, (1995).

US EPA Method 4010A. Screening for Pentachlorophenol by Immunoassay US EPA, Washington DC, January, pp. 1-17, (1995).

Vicedo, J.L., Pellín, M. (1985): Phthalates and organophosphorus compounds as cholinesterase inhibitors in fractions of industrial hexane impurities. Arch. Toxicol. 57, 46-52.

Wagner, S.L. (1983): Clinical Toxicology of Agricultural Chemicals, Noyes Data Corporation.

Wilkins, A.L., Singh-Thandi, M., Langdon, A.G. (1996): Pulp mill sourced organic compounds and sodium levels in water and sediments from the Tarawera river, New Zealand. Bull. Environ. Contam. Toxicol. 57, 437.

Zaitzev, N.A., Korolev, A.A., Baranov, I. (1990): Health-related correlation of diethylphthalate, di-n-hexyl phthalate and dialkylphthalate 810 in water. Gig. Sanit. 9, 26-28.

3 Pesticides

3.1 Measurement of Priority Metabolites Using Integrated Optoelectronic Biosensors Derived from Antibody and Synthetic Receptor Libraries

Anthony P.F. Turner[1], Andrew Rickman[2], Bertold Hock[3], Rolf D. Schmid[4], Damia Barceló[5]

[1]Cranfield Biotechnology Centre (CBC), Cranfield University, UK

[2]Bookham Technology Ltd. (BT), Oxfordshire, UK

[3]Lehrstuhl für Botanik, Technische Universität München-Weihenstephan (LB-TUM), Germany

[4]Institut für Technische Biochemie, Universität Stuttgart (ITB-US), Germany

[5]Consejo Superior de Investigaciones Cientificas (CSIS), Barcelona, Spain

Abstract. Environmental legislation and increasing public awareness of environmental issues have heightened interest in generic measurement technologies suitable for the rapid detection of a range of analytes. Of particular interest at present are analytes for which no effective conventional assay is available. These include priority metabolites and transformation products such as CIAT (6-amino-2-chloro-4-isopropylamino-s-triazine), CDET (4-acetamido-2-chloro-6-ethylamino-s-triazine) and CAAT (2-chloro-4,6-diamino-s-triazine). These, and other related compounds are of particular concern because of their high aqueous solubility. This leads to a high level of transport through pore and groundwaters to surface and seawaters. Coupled to the relatively long half lives of these compounds, this has led to their detection in the Mediterranean region due to transport via such rivers as the Ebro (Spain), Po (Italy), Rhone (France) and

Axios (Greece). These compounds are currently difficult to analyse using conventional techniques and are environmentally relevant. For these reasons, they have been selected as target analytes for this programme. The technology that will be developed, however, will be broadly applicable to analytical problems both in the environment and in process monitoring.

At present, analysis of pesticide metabolites is achieved largely via sampling and measurement in dedicated analytical laboratories using conventional analysers such as gas chromatographs and mass spectrometers. These methods are slow, require expensive instrumentation and skilled laboratory staff. Hence there is a clear need for a simpler alternative and preferably a method which allows real time monitoring in the field. The problem of analysing compounds of this type is that many of them may be present simultaneously. Hence an integrated sensor array coupled to multivariate analysis software offers a powerful and flexible tool for use in environmental monitoring.

This chapter outlines a three year programme, which began in January 1997, to develop integrated optoelectronic biosensor arrays for the detection and quantification of compounds such as CIAT, CDET and CAAT. The proposed route involves the creation of an antibody/peptide library followed by the deposition of suitable antibody fragments or molecular receptors onto a transducer structure. The output from the sensor array will then be manipulated using dedicated multivariate analysis software, with the aim of identifying and quantifying the target analytes. This type of approach has proved successful in applications such as the electronic nose and should be amenable to use in an integrated optical biosensor array. Due to the small size of these platforms, we will use silicon manufacturing technology to produce the transducers, and mass manufacturing techniques such as ink-jet printing to deposit the antibody fragments and molecular receptors. It will be possible to produce arrays of devices at low cost, which may allow disposable devices to be constructed.

Interferometry will be the primary transduction technique. This has been chosen due to its inherent extremely high sensitivity and accuracy. Furthermore, it is a flexible measurement which will enable generic devices to be produced. This is seen to be important, since it will allow the system to be adapted easily to meet changing technological and market-driven requirements. It is also possible to incorporate electrochemical sensors onto such a platform and these will be used as required.

3.1.1   Silicon-on-insulator integrated sensor plat-forms

Silicon-on-insulator (SOI) integrated optics allows the integration of phase and intensity modulators, low loss single mode waveguides compatible with fibre optics and a range of standard passive waveguide structures. Such a structure is illustrated in Figure 32. The technology also allows high efficiency coupling of laser diodes and detectors by a hybridisation technique (work conducted by Bookham Technology Ltd) and the potential integration of sensitive evanescent optical sensing elements (the feasibility of manufacturing gas sensors was demonstrated by Cranfield University and Bookham during a Department of Trade and Industry LINK programme in the UK). These attributes make this silicon-based technology an ideal cost-effective solution for integrated optical chemical/biological sensing.

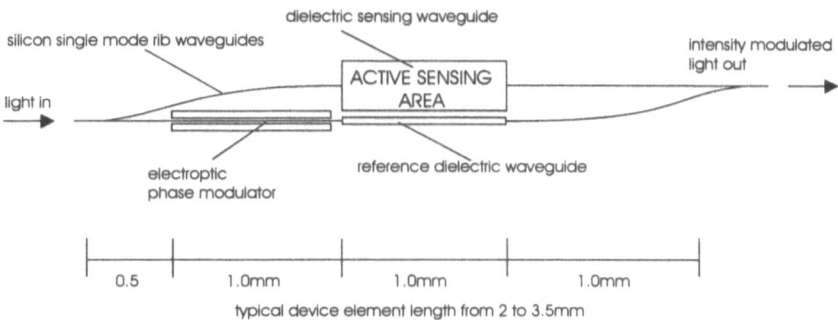

Figure 32: Active integrated sensing interferometer

The target performance characteristics include the measurement of changes in the effective refractive index of the sensor waveguide to less than 10^{-6} which is expected to equate to a concentration of the analytes of less than the order of μM. If the project objectives are successful, the sensor should be applicable to any optical evanescent field sensing application, providing low fabrication costs and high sensitivity. This project will harness this capability for the detection of priority metabolites and transformation products, but a range of other environmentally important analytes and those from other areas could be tackled in the future using this technology.

A passive (disposable) chip will contain an integrated differential path length interferometer generating interference fringes. This sensing chip will be coupled to an SOI chip for signal interrogation. It is envisaged that a final product would consist of a small instrument, possibly hand-held, and low cost disposable slides, with receptor layer arrays printed on them. The slide would be inserted into the instrument, a sample placed in a well on the slide and the instrument would provide a readout.

3.1.2   Recombinant antibodies for immunoanalysis

Quantitative immunoanalysis is usually directed towards single analytes because the assays have to rely on selective antibodies (abs). Since cross-reacting abs usually bind related analytes with widely differing affinities, the use of a single antibody for group-specific assays is usually precluded in most cases. The situation changes if array techniques can be applied (for example immunosensor arrays). Here, two extreme cases can be considered:

- Multianalyte analysis with selective abs which recognise analytes with entirely different structures.
- Multianalyte analysis with cross-reacting abs, which bind analytes of related structure.

In the latter case, chemometric approaches can be used for the quantitative analysis of single compounds.

Present technology is limited by the time and costs associated with the development of new abs. Conventional ab production relies on a tedious immunisation process. Under favourable circumstances, new polyclonal abs can be produced after approximately 3 months and new monoclonal abs after about 6 months. If the immunisation fails, the whole process needs to be restarted.

The present situation is likely to change because of recombinant (rec) ab technology. A small number of laboratories have reported the first rec abs for pesticides and toxins. For example, A. Karu (US) reported rec abs for diuron, B. Hammock (US) for atrazine and M. Morgan (UK) for mycotoxins. In addition, LB-TUM has already developed monoclonal antibodies (mabs) for a range of compounds, e.g. Hg^{2+} ions, 3,4-dichlorophenol, as well as for surface antigens of soil fungi and bacteria. More recently, LB-TUM has succeeded in preparing recombinant single-chain (rsc-ab) and Fab (r-Fab) fragments from hybridomas producing mabs selective for atrazine and terbuthylazine. LB-TUM is thus fully equipped for the preparation of both hybridoma and recombinant antibody fragments for environmental applications. Some of these hybridoma cell lines have been subsequently cultured by ITB-US to produce and purify mabs at the gram scale.

Earlier experience in LB-TUM has shown that r-Fabs are more stable and easier to produce than rsc-abs. However, rsc-abs are more convenient for the screening step: a simple "phage display" method can be used, where the phage particles present sc-abs on their surface. DNA coding for the variable region of the abs can be recovered from the phages and exploited for subsequent work. LB-TUM have obtained rec ab fragments for atrazine assays, which were synthesised starting with mRNA from existing hybridoma cell lines, and compared favourably with the original monoclonal antibodies. In order to extend the selectivity and affinity of rec ab fragments, it is proposed that recombinant clones are used for the construction of new DNA libraries. It is expected that a multitude of new rec ab selectivities and affinities will be derived within a group of compounds, such as priority metabolites and transformation products.

Two different strategies will be applied:

- Ab fragment coding sequences will be cloned into suitable vectors and multiplied in bacterial strains with high mutation rates.
- A modification of DNA sequences coding for the binding sites of rec ab fragments can   be achieved by PCR using primers which produce errors during DNA amplification.

Large libraries (eg. phage display libraries) can be selected by columns coated with suitable target conjugates.

The main benefit of these techniques is seen in the field of multivariate analysis of closely related compounds. However, entirely new ab selectivities may be obtained from ab libraries derived from spleen cells of non-immunised animals. A further advantage of the rec ab technique, is related to the possibility of constructing fusion products carrying functional groups for the immobilisation of rec ab fragments

on sensor surfaces. Thus, the rec ab fragments can be properly oriented for optimal analyte binding.

The monoclonal and the recombinant approach are compared in Figure 33.

It is proposed that work will start with available rsc-abs selective for atrazine and for terbuthylazine. From these, mutant antibodies will be prepared by PCR techniques and these mutants will be screened for high affinity rsc-abs toward priority one metabolites by an immunoaffinity screening procedure, developed by LB-TUM. Immunoconjugates (e.g. priority one atrazine and terbuthylazine metabolites such as deethylatrazine or deisopropylatrazine) are immobilized on the surface of sepharose beads and serve to bind and characterize phages expressing rsc-abs of high affinity. The variable regions from single chain fragment DNA (scFv-DNA) are then cloned into suitable vectors carrying the DNA segments for the constant domains of Fab fragments. As a final step, r-Fab fragments are expressed.

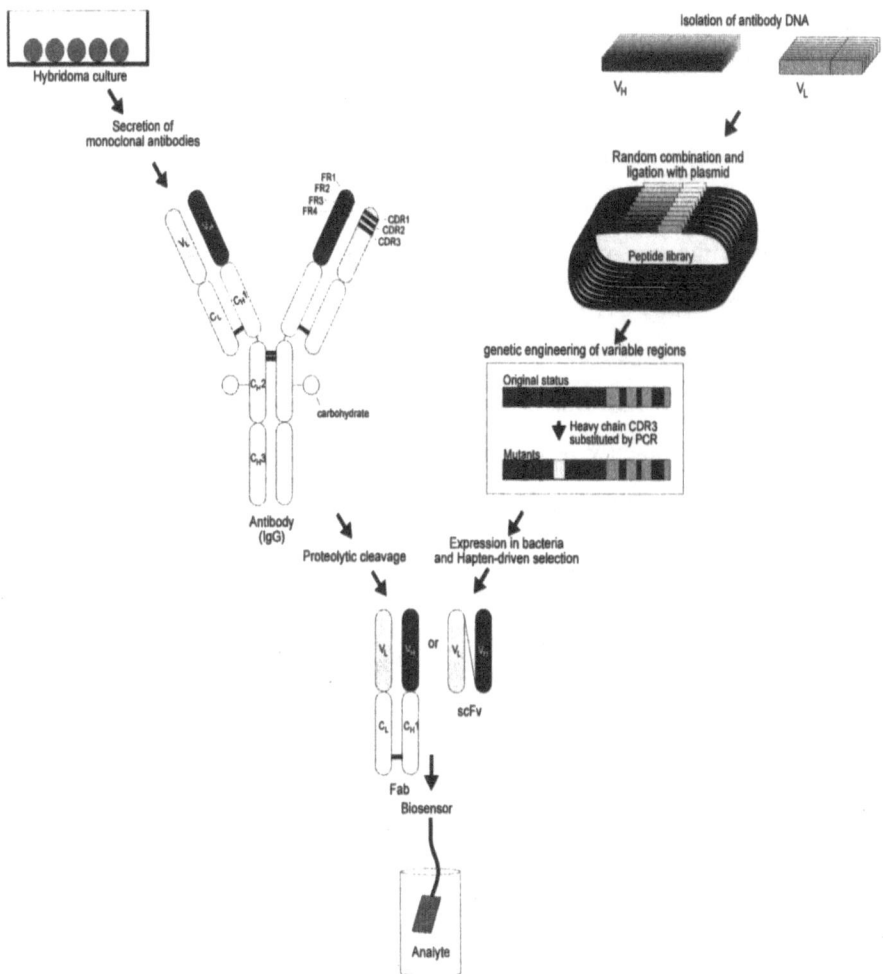

Figure 33: Comparison of the monoclonal and the recombinant approach for the production of antibodies

In addition, synthetic libraries will be exploited for the further improvement of existing antibody fragments. This work will be done in collaboration with CBC, where expertise on the synthesis and screening of synthetic peptide libraries is available (see next section). LB-TUM will start with pre-optimised phage clones, which are identified by the methods mentioned above. DNA coding for the variable region of the heavy chain is randomly combined with the corresponding region of light chains where CDR regions are replaced by segments taken from the DNA library. The phage display method is used again to screen for suitable ab properties. As before, the final product will be r-Fabs of high affinity for priority one metabolites.

ITB-US, having equipment and skills in computer modelling of proteins, will build structure models of the mutant sc-abs and r-Fabs thus obtained, and also prepare putative models of the sites of hapten-antibody interaction, suggesting preferred amino acid side-chains for further mutation studies ("protein engineering cycle"). ITB-US will also produce and refold optimized r-Fabs in the larger scale and supply samples to Participants 1-3. ITB-US has a suitable host-vector system available and has already prepared mg quantities of active recombinant Fab fragments of a terbuthylazine-specific antibody whose structure gene originated from LB-TUM.

3.1.3 Synthetic peptide libraries

Molecular recognition, the specific interaction of substrate and receptor is central to the study of intermolecular forces, the mimicry of biological systems and the development of supramolecular devices. The traditional "lock and key" theory is a good model for molecular recognition. This process embraces the "lock", the "key" and the "fit". The lock refers to the crevice inside or on the surface of a molecular entity (biopolymer, pharmacophore, crystal or host) accommodating the key: a whole molecule or part of it (ligand, pharmacon, asymmetric unit or guest). The term "fit" stands for at least three types of interactions: steric, electrostatic and hydrophobic. Computer modelling offers an efficient tool for the representation of various aspects of recognition.

The development of artificial receptors for various purposes remains an important synthetic challenge. Most synthetic molecular receptors fall into two broad classes: those that bind molecules in aqueous solution and those that bind in relatively non-competitive organic solvents. Among various molecular clefts and cavities, few synthetic receptors have been developed to bind substrates in aqueous solution. Molecular receptors that bind substrates in water generally achieve binding with the aid of hydrophobic interactions in conjunction with hydrogen bonding, resulting in recognition.

We propose to design new types of water soluble receptors which form salts with the triazines and also form strong hydrogen bonds with the triazine ring system. The hydrogen-bonding groups will be incorporated inside the macrocyclic molecules to shield the hydrogen bonds from water to achieve specific binding.

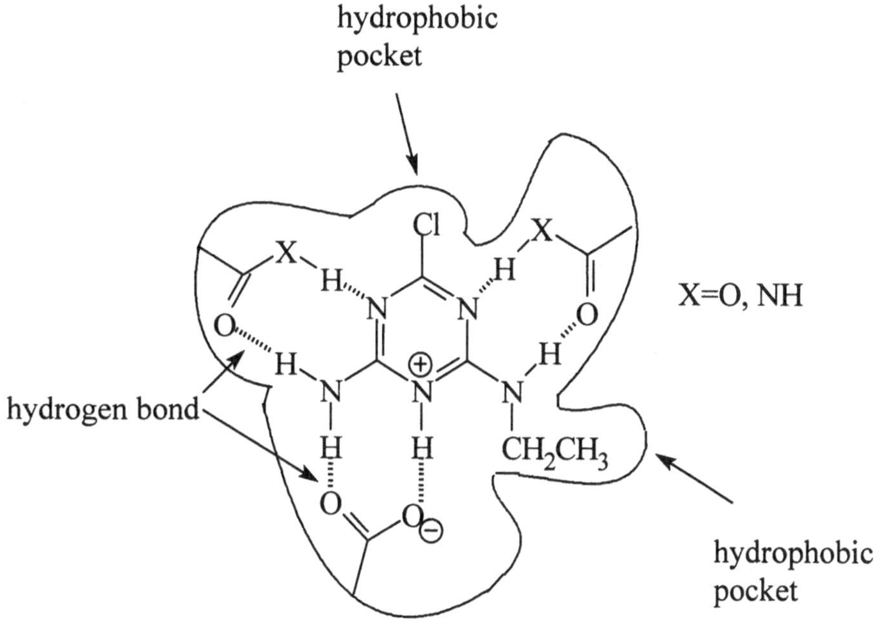

Figure 34: The molecular receptor model for CEAT

Compared to their parent compounds, the primary metabolites, CEAT, CIAT and CAAT have triazine rings with one or more amino groups and are more water soluble, leading to poor recovery by conventional methods. They are weak bases and form salts less easily and hydrogen bonds more easily than their parent compounds. When they bind to their receptors, they request smaller hydrophobic pockets and different electrostatics. These are the chemical principles guiding the design of the receptor (Figure 34).

Molecular modelling will be extensively used for the design of the molecular receptor. The modelling software Quanta and Charm package will be used to study the properties of substrates (CEAT, CIAT, CAAT) and interactions with manually created molecular receptors to manipulate the size of hydrophobic pockets and the macrocyclic ring and to guide the selection of the building blocks for the polar macrocyclic receptor library. This work will be carried out jointly by ITB-US and CBC.

Combinatorial synthesis, which is also referred to as molecular diversity, is one of the most exciting and rapidly growing areas in ligand discovery. It may be defined as the systematic and repetitive, covalent connection of a set of different "building blocks" of varying structures to each other to yield a large array of diverse molecular entities in many combinations using chemical, biological or biosynthetic procedures. These intentionally created libraries can be screened for biological activity in a variety of different formats (eg. libraries of soluble molecules; libraries of compounds tethered to resin beads, silica or other supports; recombinant peptide libraries on bacteriophage and other biological display vectors). Theoretically, the number of possible individual compounds (N), prepared by an ideal combinatorial synthesis is determined by two factors: the number of building blocks for each step (b) and the number of synthetic steps in the reaction scheme (x). If an equal number of building blocks are used in each reaction step, then $N = b^x$. Among various methods for the construction of chemical libraries, the resin-based mix and split technology is the most popular. This is illustrated in Figure 35.

Through a protocol of separating and mixing beads during the synthesis, each bead in the final library has a product from a single, specific reaction sequence chemically bound to it and that product is likely to differ from that bound to another bead. After selecting a particular bead which has some desirable property, the identity of the attached compound is determined by analytical chemistry.

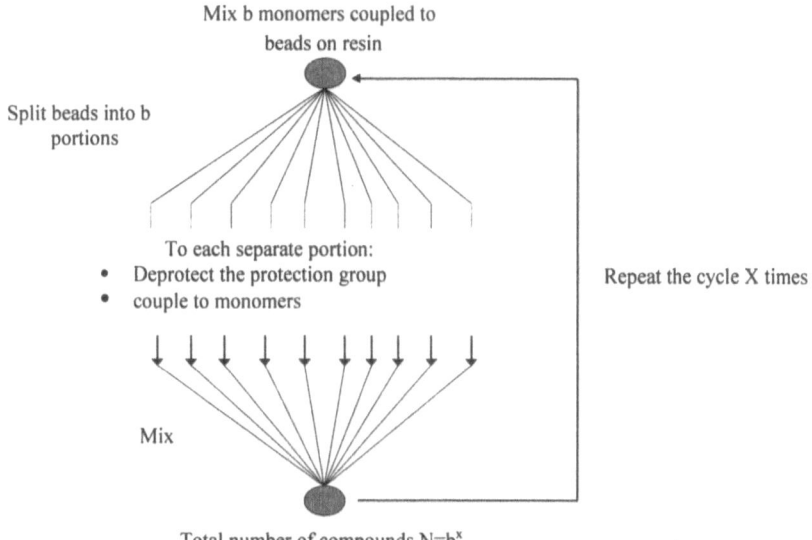

Mix b monomers coupled to
beads on resin

Split beads into b
portions

To each separate portion:
- Deprotect the protection group
- couple to monomers

Repeat the cycle X times

Mix

Total number of compounds $N=b^x$

Figure 35: Split and mix technique in combinatorial synthesis

Building blocks for the library will be chosen according to the modelling study, which will include polar monomers such as aspartic acid and serine and hydrophobic spacers such as alkyl chains and/or aromatic rings. Both natural amino acids and functionalised amino acids could be used as building blocks. The library will be constructed using the mix and split technology to produce an initial library of 10^6 compounds to obtain a reasonable size and good structural diversity. Ten monomers bound to the resin will be evenly mixed and split into ten pools. Ten pools containing a total of 10^6 compounds will be obtained after six rounds of reactions with mixing and splitting at the end of each cycle.

Polystyrene beads have been used in the construction and screening of support-tethered libraries. The bead material is available in a variety of diameters and with several different functional groups (eg. NH_2, OH, SH, Br, COOH). The 90 μm diameter beads give rise to approximately 2.9×10^9 beads/g resin, with a loading of 80-100 pmol per bead. Each bead should provide adequate material for direct microanalysis after selection by the binding assay. Two types of beads are proposed as supporting materials: NH_2-Merrifield resin and Wang cross-linked polystyrene resin. The former will be used to synthesise the peptide library tethered to the resin, while the latter will be used to resynthesise the library once active compounds have been identified, since the resin can be cleaved easily.

The synthesised library will be screened in each pool to find an effective receptor for each analyte using standard methods. The beads which bind to a substrate can be picked out under a microscope by fluorescent labelling or radio-labelling the substrate. The library will also be screened against other substrates in parallel to obtain a specific receptor for each.

The structure of the most effective receptor will be elucidated by analytical methods following resynthesis of the library. The library will then be subdivided into a smaller pool and resynthesised until the pool containing the desired receptor is obtained. This will then be carried out on a larger scale to obtain sufficient material to allow its full characterisation and investigation of its applications. Finally, computer modelling will be used in the further modification of the receptor.

Thus, two powerful techniques of modelling and combinatorial synthesis can be used in parallel to lead rapidly to effective receptors. In combination with the rec-ab library, as described earlier, we will have a large range of receptors available for use in the biosensor array.

3.1.4   Biosensor assay design

There are many potential assay formats which could be used. The two most common conventional formats are sandwich and displacement assays. One of these may be suitable for the purposes of this project. However, it may be necessary to consider amplification formats, or possibly the use of an infra-red (IR) active label, which could be detected at the proposed wavelengths of approximately 1200 nm. ITB-US and CBC will design suitable assays based on these techniques and will collaborate with CSIS to characterise and optimise the system adopted.

One of the major problems which affects devices of this type is that of non-specific effects, such as the binding of various proteins from the sample to the surface of the sensor. One way of overcoming (or certainly reducing) this effect is to use a reference sensor arm. This can be achieved simply by fabricating an optical reference arm with no active receptor layer. Alternatively, arrays of sensors can be constructed and the signal due to the binding of each analyte can be calculated using multivariate analysis techniques. This has the further advantage that interference from similar compounds can be minimised if the receptor specificity is not sufficient.

A further problem which afflicts many immunosensors is that of poorly controlled ab orientation. As mentioned earlier, we will take steps to minimise this effect by incorporating fusion products incorporating functional groups to aid the immobilisation process onto the sensor surface.

3.1.5 Antibody fragment and molecular receptor deposition

Many automated production technologies are currently available for the construction of biosensor devices. Of these, printing processes, such as screen-printing, ink-jet printing and air-brush printing have been the subject of a number of studies aimed towards the deposition of biological and chemical components such as antibodies and molecular receptors.

Due to the small volumes of reagents and high degree of accuracy that will be required for this application, it is proposed that ink-jet printing will be used to deposit suitable reagent layers on the surface of the sensor array. The ink-jet process is most familiar in the form of the office-based ink-jet printer. A similar technology is very widely used to print, for example, sell-by dates on food packaging. A diagram of such a print head is shown in Figure 36.

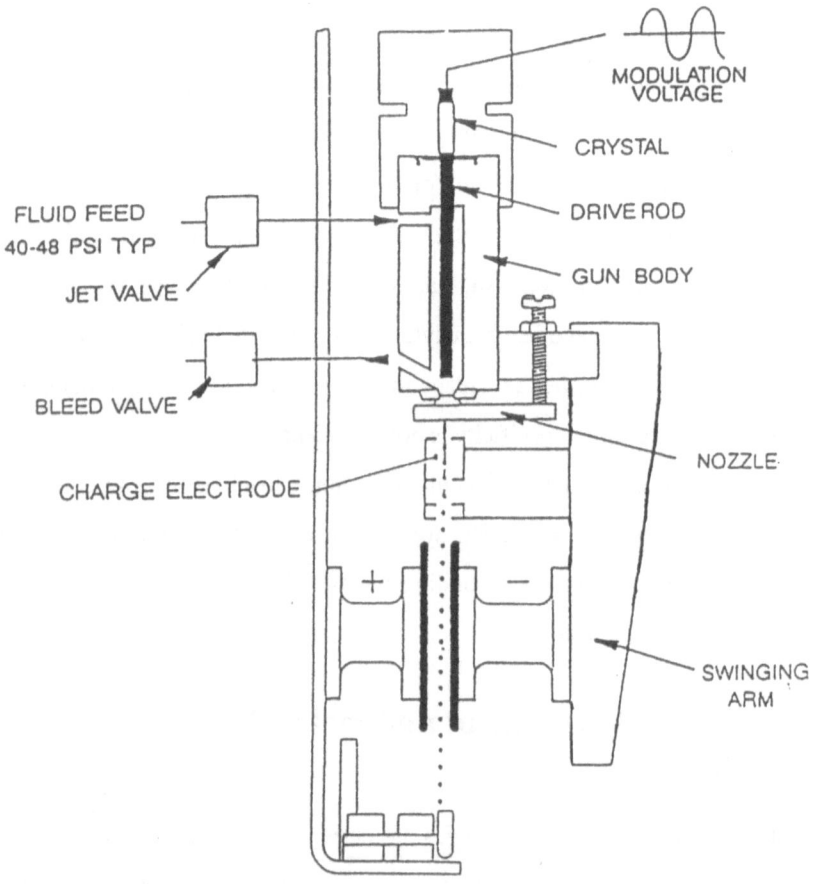

Figure 36: Diagram of ink-jet printhead

The process operates by circulating a solution through a small hole (typically 35-75 pm diameter) in a nozzle under a pressure of approximately 3 bar. Shock waves are sent through the fluid via the drive rod, which breaks the jet into individual droplets at a rate of approximately 64 kHz. The jet is aligned such that it enters the return tube for recirculation. Printing is achieved by charging individual droplets, under the control of a microprocessor, using the charging electrode. The charged drops are then deflected on passing through two parallel high voltage deflector plates. A pattern is hence built up in dot-matrix form.

Due to the microprocessor control, it is extremely simple to change the print pattern and the volume of reagent dispensed with such a printer. It is also a non-contact process which means that it is suitable for deposition of material on non-planar and delicate surfaces, should the need arise. The solution must, however, be electrically conducting (so that the droplets can be charged) and free of large particulates. The material is only deposited where it is required and due to the small size (in the order of 1 nl) of each droplet, the process is economical with reagents.

A simplified print arrangement is currently under development at BioDot, a company which is working with Cranfield Biotechnology Centre to develop printing technologies for biological applications. The Biojet instrument will be available for this project and it is proposed that this will be evaluated and compared with the ink-jet process.

3.1.6　Multivariate analysis software

Pattern recognition is the term used to describe the methodology of solving classification-type problems traditionally encountered in the physical and engineering sciences. Recent advances in chemical sensor and biosensor technology have led to arrays of devices, thus extending the analytical application potential of pattern recognition to the analytical sciences.

There are two main approaches to pattern recognition: parametric and non-parametric. The former relies on estimating or obtaining the probability density function of the parameters used to characterise the response of a system, whereas the latter requires no assumption about the fundamental statistical distributions of the data. Commercially available packages are currently being introduced (such as the "Pirouette" package produced by Infometrix) and application specific software is currently under development at Cranfield University. Adaptation of this technology will be undertaken in order to obtain the most suitable package for the determination of environmentally relevant analytes, using a sensor array.

3.1.7   Sensor validation

Priority metabolites and transformation products are currently analysed by extraction from the aqueous phase using dichloromethane liquid-liquid extraction (LLE) or solid-phase extraction (SPE) with either cartridges or Empore discs followed by gas chromatography - nitrogen phosphorus or mass spectrometric detection (GC-NPD or GC-MS respectively). These methods, although robust and well established, are time consuming, expensive and require expensive instrumentation. These methods usually offer good recovery of, for example, the parent triazine compounds, but poor recoveries of metabolites which are less hydrophobic. This is mainly due to the fact that large water volumes (typically 1-2 litres) are required and as a consequence, early breakthrough of the polar metabolites occurs. For instance, typical recovery of deisopropylatrazine varies between 17-36%.

Another approach is to use automated on-line SPE followed by liquid chromatography (LC) with either diode array or mass spectrometric determinations (LC-DAD or LC-MS). In this way, water volumes of 100-150 ml are required and as a consequence, the major problem of early breakthrough encountered with off-line SPE can be avoided. In this instance, the recovery of deisopropylatrazine can be improved to about 65%. Problems occur due to the type of water matrix. When relatively clean water matrices are analysed (eg. drinking or ground water), then the polar metabolites can be determined with on-line SPE-LC-DAD by preconcentrating a low water volume of 50-100 ml. However, when water matrices are relatively dirty (eg. river, esturine or waste water), then problems occur related to interferences in the chromatographic analysis of the humic substances, which do not allow the determination of polar metabolites. In addition, binding of the analyte with these interferent substances occurs, making accurate determination at trace levels difficult.

Enzyme-linked immunosorbent assays (ELISA) for different triazines and triazine metabolites have been the most popular assays used in environmental analysis. These are a relatively inexpensive alternative to GC and LC methods and good agreement between the methods has been observed.

Substances present in water and soil affect ELISA determinations. For example, false positive results can be caused by structurally similar compounds. Other sources of false positive responses are synthetic or natural substances, such as dissolved organic carbon (DOC), which may interact weakly with the antibody, but may be present at such a high concentration, that a relatively strong signal results. These problems may be reduced by using a sample preparation step such as SPE or LLE followed by a clean-up step (LLE-Florisil).

Validation studies will involve a number of stages:

- Evaluation of antibody fragments and molecular receptors will be carried out using ELISA type technology. These will be tested for the analyte of interest as well as known, common interferents such as humic substances. The effect of sample salinity will also be assessed.

- Comparison of ELISA methods versus automated on-line SPE chromatographic methods, mass spectrometry confirmation (to avoid false positives). Evaluation of matrix interferences.

- Validation of the biosensor developed, by automated chromatography/mass spectrometric methods. The influence of all parameters previously evaluated for ELISA will be studied.

- Analysis of inter-laboratory samples and natural water samples from different origins. Evaluation of the biosensor under field conditions.

The first antibody fragments for atrazine from ITB-US have shown an IC_{50} of 5 ppb with a limit of detection of 0.1 ppb. Further fragments are currently being evaluated.

3.1.8   Conclusions

The technologies outlined above will initially provide a means of monitoring three environmentally relevant analytes. At present, these are difficult and time consuming to measure. The current procedure involves sampling, followed by analysis in an analytical laboratory using expensive and complicated equipment. It is proposed that small, simple to use and inexpensive sensors could be constructed. This will allow measurements to be made quickly and at low cost. A hand-held instrument would be possible, which could allow real-time readings to be made in the field. This would greatly assist users, which would include Governmental Environment Officers, National River Authorities and Regional Water Boards as well as Environmental Consultants.

Improvements in the monitoring method would allow better assessments of the impact of factors such as changes in farming practise and climatic fluctuations on the levels and distribution of hazardous compounds. Monitoring and legislation can then be used together to better safeguard the European Community.

Many claims have been made about the potential size of the market and the possible impact of biosensors. These estimates vary widely, but are all in agreement that it is very large. At present, the world market for commercial devices is dominated by the medical area. This is not perceived to change in the foreseeable future. However, the use of biosensors in other areas, such as in environmental and process applications is widely predicted to increase rapidly. Advances, such as those proposed in this project together with greater demands for information due to legislative changes and increased public awareness will speed up this process.

3.1.9   Acknowledgement

The support of DGXII, EC Climate and Environment for this work is gratefully acknowledged.

3.2 River Analyser - Multiresidue Immunoanalytical Monitoring Tools

Andreas Brecht[1], Claudia Barzen[1], Albrecht Klotz[1], Günter Gauglitz[1], Richard Harris[2], Geoffrey Quigley[2], James Wilkinson[2], Soizic Fraval[3], Pascale Sztajnbok[3], Damia Barceló[4], Jordi Gascón[4], Michael Steinwand[5], Ram Abuknesha[6]

[1]Institute of Physical Chemistry, University of Tübingen, 72076 Tübingen, FRG

[2]Optoelectronics Research Centre, University of Southampton, Southampton SO17 1BJ, Great Britain

[3]Anjou Recherche / Compagnie Générale des Eaux, Chemin de la Digue BP 76, 78603 Maisons-Laffitte, France

[4]Centro De Investigacion Y Desarrollo, Jordi Girona 18 - 26, 08034 Barcelona, Spain

[5]Perkin-Elmer GmbH, Bodenseewerk, 88647 Überlingen, FRG

[6]GEC Marconi Materials Technology, Elstree Way, Borehamwood, Hertfordshire WD6 1RX, Great Britain

Abstract. Immunoanalytical techniques find growing acceptance in the field of environmental monitoring (Van Emon and Gerlach 1995). Although immunoassays have many attractive features, the lack of multiresidue approaches is a serious obstacle in environmental monitoring. In many environmental situations the simultaneous determination of more than one analyte is required. The situation is further complicated by the fact, that the analyte panels, which are to be detected, may vary with location and season.

The RIANA project (RIver ANAlyser) is focused on immunoanalytical tools, that allow to detect multiple analytes in a single run. The application areas foreseen are online monitoring - e.g. at pumping sites or transportable devices for field use. In this paper we describe the concept and give first results.

3.2.1 Introduction

Rivers, throughout Europe, form a major source of process water as well as raw water for human consumption. European rivers, however, collect a significant proportion of both treated and untreated waste water from sewage plants and industro-agricultural processes. This, as well as the use of rivers as waterways for transportation, yields to a very complex environment which is prone to pollution. This environment is subsequently transferred to the estuarine and coastal sea areas. The situation is aggravated in basins with reduced water exchange such as the Mediterranean and the Baltic Seas where pollution by rivers can cause serious environmental problems affecting human health and the economical prospects of coastal activities.

The role of rivers as a vital source of drinking water as well as for industrial/ agricultural processes necessitates continuous monitoring and control of the water quality (Meulenberg and Stoks 1995). This monitoring should be effected at pumping sites, industrial outputs and various important interfaces. In essence, the health of major rivers should be monitored continuously and regularly from source to coast. This implies that concentrations and transition/ life-cycles of pollutants in rivers should be monitored and effective control established.

The objective of the RIANA project is to develop a system for monitoring organic compounds present in polluted river water and in surface water. The system will be a multiresidue bioanalytical device, allowing to test in one sample a multitude of simultaneously present agents. The system envisaged is expected to be more cost effective than established reference techniques, to allow the detection of a broader range of analytes with a single device than with established reference techniques, and to avoid most of the sample pre-treatment common in current environmental multiresidue analysis with concomitant advantages in speed. The multianalyte approach pursued will overcome one of the major limitations of immunoanalytical techniques in environment - the limitation of a single analyte per test. An overview of the main building blocks is given in Figure 37.

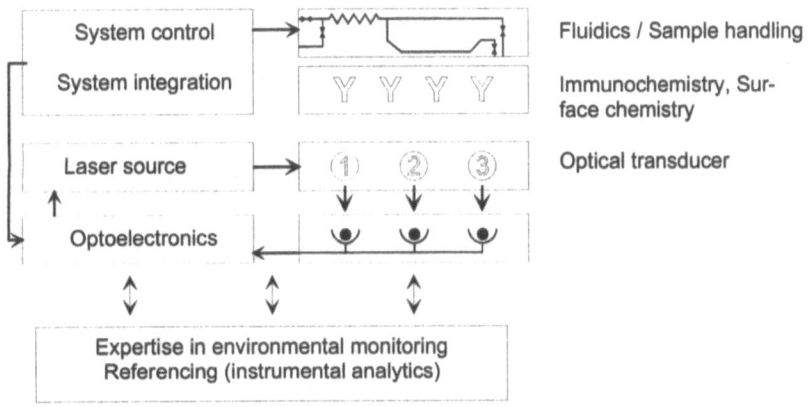

Figure 37:   Building blocks of the RIANA approach. Major building blocks and research topics are immunochemistry, surface chemistry, and the optical transducer. Fluidics, optoelectronics and system control serve to "glue" the system together. Referencing to established techniques is a central part of the project

3.2.2   Materials, methods and concepts

3.2.2.1   Transducer

The transducer system is based on fluorescence detection (Wadkins et al. 1995). Targets are effective excitation and spatially resolved detection of fluorescence labelled antibodies bound to the transducer surface. Initially a simple bulk-optical device is used. This will be replaced by an integrated optical transducer element which will lead to reduced active area and improved excitation and detection. For each analyte to be detected, a defined spot of the transducer surface is modified with a

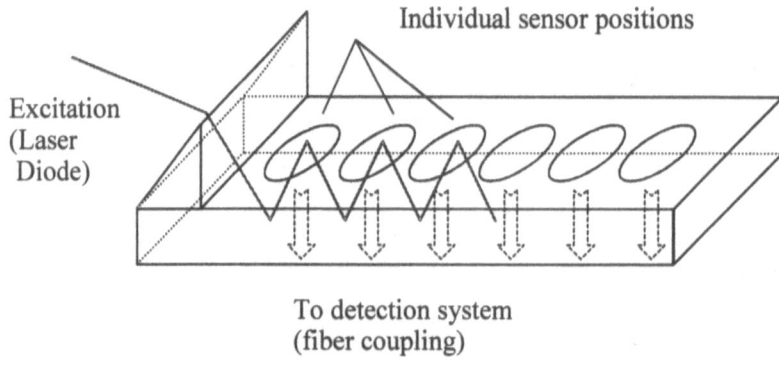

Individual sensor positions

Excitation
(Laser
Diode)

To detection system
(fiber coupling)

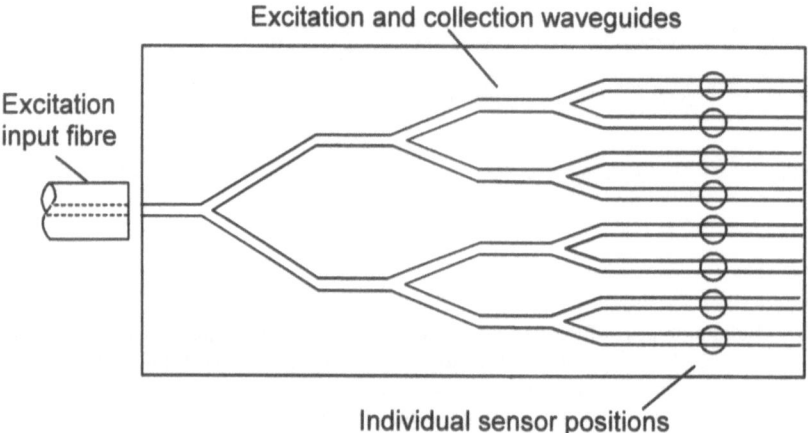

Excitation and collection waveguides

Excitation
input fibre

Individual sensor positions

Figure 38: Optical multispot transducers investigated within the RIANA project. Top: simple bulk optical transducer - detection spots are about 1.5 mm * 1.5 mm in size; Bottom: integrated optical channel waveguide device - detection spots are about 0.1 mm * 5 mm in size

derivate of the respective analyte (Figure 38). Transducer design starts with numerical modelling of different structures, starting from monomode waveguides, modified by high index cover layers to optimise fluorescence excitation. The waveguide model developed to analyse the first generation integrated optical (IO) sensor uses a transfer matrix approach to allow the analysis of multilayer structures. The argument principle method is used to determine the number and effective indices of modes which exist in the multilayer region of the sensor. A more detailed description of the model is given by Harris and Wilkinson (1995).

3.2.2.2 Immunochemistry

Antibodies will be used to bind selectively to the analytes of choice. Antibodies will further be used as components of a flexible immobilisation system, that allows the selection of the transducer surface on-line during the assay (Brecht and Abuknesha 1995, Abuknesha and Brecht 1997). The performance of immunochemistry is assessed in a conventional ELISA testbed (Tijssen 1985).

Table 18: Target analytes selected for the RIANA project

Toluene	Pentachlorophenol
2,4,5-Trichlorophenol	2,4,6-Trichlorophenol
Hydroxycarbofuran	Desethylatrazine
Atrazine	Simazine
2,4-D	Methylchlorophenoxypropionic acid
Alachlor	Glyphosate
Isoproturon	Nonylphenol

For the detection of analytes a competitive test scheme will be implemented. ELISA results will serve as a reference. Competition will occur between the free analyte and compounds of the biochemical immobilisation system conjugated with analyte derivatives. The detection will occur in a heterogeneous format, i. e. at the transducer surface. The analytes addressed within the project cover pesticides as well as industrial pollutants (Table 18). By purpose a set of compounds was selected, which can not be determined within a single chromatographic run by conventional

152

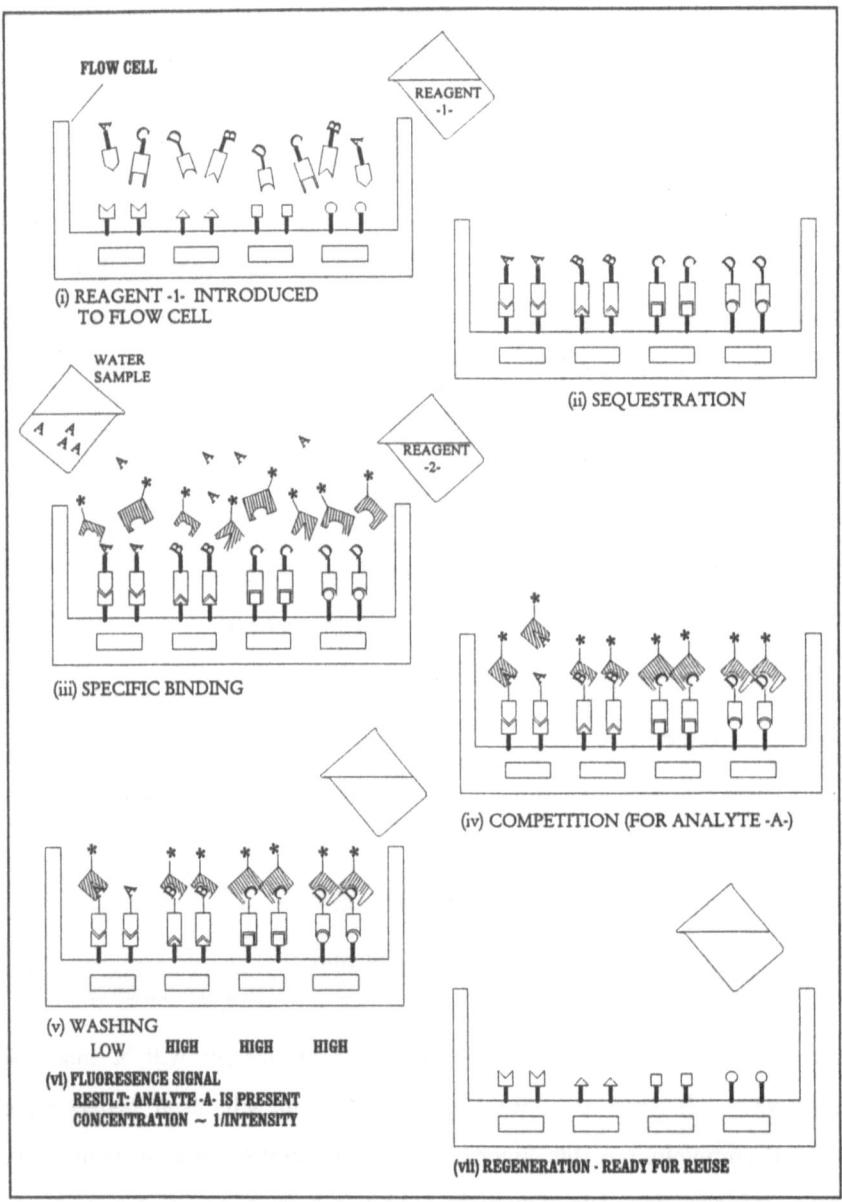

Figure 39: Assay concept. Antibody conjugates serve as "auxiliary immobilising system" and bring
analyte specific chemistry to selected areas at the transducer surface. Sequential and non-
sequential assay concepts are possible

instrumental analysis. Also polar and acidic compounds were included, which may escape typical extraction procedures (Chiron and Barceló 1993).

Immunogens were made by coupling selected analyte and auxiliary ligand derivatives to mixtures of bovine serum albumin, key hole limpet haemocyanine and fetuin. Antibody activity in the various antisera was assessed by antibody dilution response curve experiments in which immobilised hapten-protein conjugates were used to assess antibody titre in an ELISA format.

An auxiliary immobilising system consisting of antibodies specific for haptens without environmental relevance will serve to target the detection compounds to selected areas of the transducer surface (Figure 39). The spatially resolved immobilisation of the haptens of the auxiliary system is achieved by inkjet type microdrop delivery systems (Blanchard et al. 1996).

Table 19: Reference sample matrix used in the RIANA project

Parameter	Value
pH	7.5
Conductivity (mS/m)	50
Total Hardness (mg CaCO3 / l)	200
Chloride (mg/l Cl)	50
Sulphates (mg/l SO4)	40
Silica (mg/l SiO2)	10
Magnesium (mg/l)	20
Potassium (mg/l)	10
Sodium (mg/l)	50
Nitrates (mg/l)	20
Humic acids (mg/l)	4

3.2.2.3   Reference analysis and validation

Validation is carried out by laboratories with a high degree of experience in analysing complex environmental samples. Where standardised methods do not exist neither at

154

the national level nor at the European level, US-EPA protocols will be adapted (Marco et al. 1995). A first step in referencing the RIANA tools is comparative studies with ELISA type assays developed with the RIANA immunochemistry. Later on the RIANA analyser hardware will be compared to instrumental analytical techniques. Real world samples from sampling campaigns will be used. Initial trials and calibrations are carried out in a sample matrix matched to values typical for European water samples (Table 19).

3.2.3   Results and discussion

3.2.3.1   Transducer

The efficiency of fluorescence excitation of bulk optical and integrated optical structures can be significantly improved by the introduction of thin high index cover

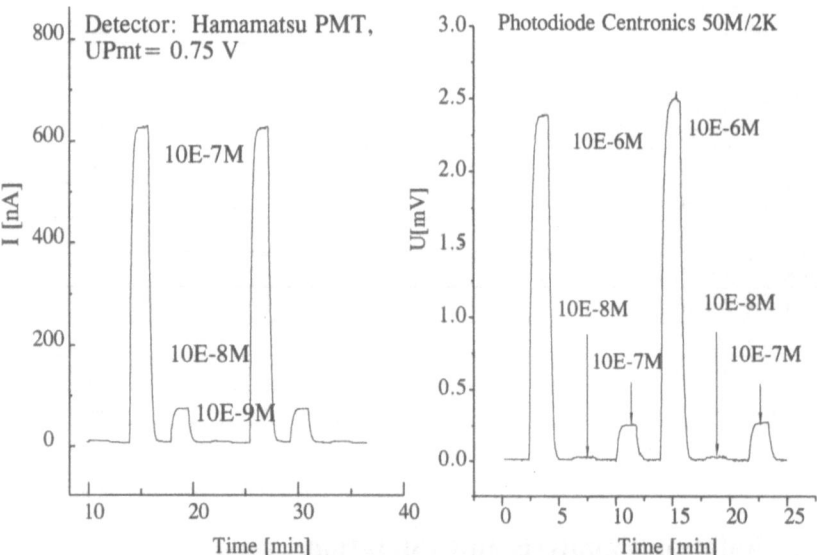

Figure 40:   Detection of fluorophore solutions with bulk optical transducer. Concentrations of 10^{-8}M can be detected with both, photomultiplier and solid state detectors. This corresponds to 0.01% of a monolayer of fluorescence labelled antibodies. The photomultiplier is one order of magnitude more sensitive than the photodiode

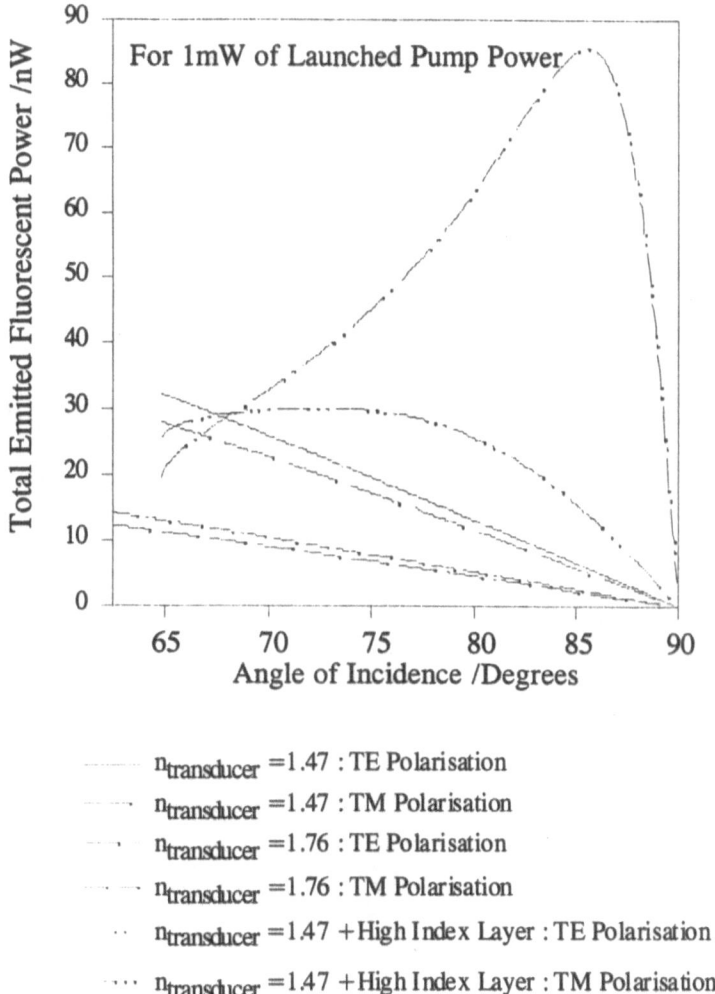

For 1mW of Launched Pump Power

Figure 41: Total emitted fluorescent power from a monolayer of dye-modified antibodies adsorbed to the surface of various bulk optical sensor structures

layers on top of the substrate or waveguide. The sensitivity of bulkoptical sensor structures, with and without high-index films, is plotted in Figure 41. Model calculations indicate similar behaviour for monomode channel waveguide designs. High index coated waveguide transducers were manufactured and are currently being tested. During a test cycle the formation of up to 1% of an antibody monolayer is expected at the transducer surface. This surface coverage can be converted to a corresponding bulk fluorophore concentration, indicating, that the transducer should be capable of detecting 10^{-8}M fluorophore solutions. This could be demonstrated for a long wavelength excitable dye (Amersham Cy5.5) with the bulk optical transducer (Figure 40).

3.2.3.2 Immunochemistry

Titration tests have demonstrated detectable antibody activity to analyte derivatives in all samples tested. The detectable antibody binding activity levels showed considerable variation with different structures. As expected, antisera to small analytes

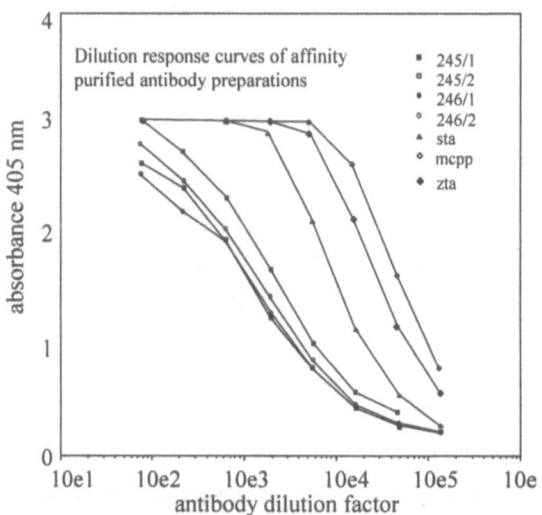

Figure 42: Antibody dilution response curves of affinity purified antibody preparations. The tests were carried out by ELISA using antibody stocks of 10 mg/ml. The graphs show the substantial differences between the antibody titres of the various antibody preparations. 2,4,5 = 2,4,5-Trichlorophenol; 2,4,6 = 2,4,6-Trichlorophenol; mcpp = Methylchlorophenoxy--propionicacid; sta = Simazine-thioaceticacid; zta = atrazine-thioaceticacid

such as toluene, glyphosate and nonylphenol showed lower levels of activities than the larger chlorophenol haptens (Figure 42). Antisera to auxiliary ligands showed much greater levels of binding activities by comparison to the responses to the panel of analytes. This is due to the relatively larger molecular weights of the selected auxiliary ligands and more antigenic structural features of these substances.

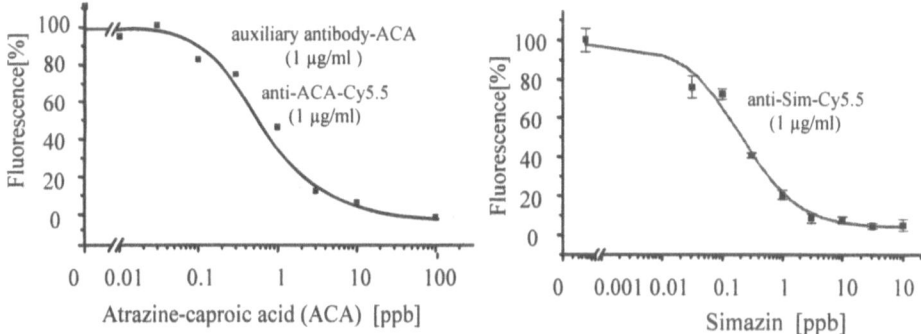

Figure 43: Proof of principle for auxiliary immobilising system. Left: calibration for a s-triazine derivative with auxiliary system. Right: calibration for a s-triazine derivative without auxiliary system

The bulk optical transducer was used to implement first fluoro-immunoassays. The transducer surface was modified with an amino-dextran layer to reduce non specific binding. To this layer either an analyte derivate or an auxiliary hapten was coupled by carbodiimide chemistry (Piehler et al. 1996). Antibody concentrations were adjusted between 0.5 µg/ml - 1 µg/ml. Analyte specific antibodies were labelled by Cy5.5 with a ratio of about 2 fluorophores per antibody molecule. A flow injection analyser was used to deliver the sample solution to the transducer surface. Increasing amounts of free analyte were added to suppress the antibody binding to the surface. With both systems calibration curves were obtained (Figure 43). Individual assays could be run within 10 - 15 minutes. In both cases test midpoints below 1 ppb could be reached. It must be noted, that s-triazine conjugates represent relatively good immunogens. The detection with the bulk optical transducer at the concentrations

158

selected (0.5 - 1 µg/ml antibody) was possible at a signal to noise ratio of about 100. This is sufficient, but further reduction of antibody concentrations would require improved detection systems.

3.2.3.3   Reference analysis and validation

The ELISA type assays are a first tool for assessment of immunochemistry performance. An important parameter is the cross reactivity pattern obtained for different antibody preparations. In Figure 44 the cross reactivity of an anti-atrazine antibody against two other s-triazines is shown. Not unexpectedly cross reactivity is considerable. The test midpoints are between 1 and 10 ppb.

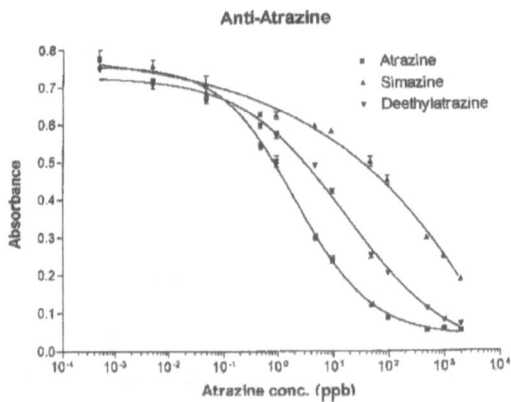

Figure 44: Assessment of crossreactivity for analyte specific antibody

3.2.4   Outlook

The first results of the RIANA project are encouraging. Immune responses against a wide range of environmentally relevant substances were obtained. The transfer of results from ELISA to the optical transducer works. With high affinity antibodies test midpoints below 1 ppb can be reached. Improved optical detection systems will allow the further optimisation of assay performance.

The use of ELISA type assays as a testbed for the assessment of immuno-chemistry performance is a pragmatic, and effective approach. The data will not directly scale to the response from the optosensor setup, but the general tendency of performance can be estimated.

The immune responses obtained for various antigens indicate, that it will be difficult to obtain this high level of performance for all the analytes. The careful optimisation of the hapten used for production of the auxiliary tracer conjugate may allow to do some optimisation in these cases. Also cross reactivity may preclude the individual substance identification in some cases. Sum parameters however should be readily obtainable.

So far the auxiliary immobilising system could be shown to work. However, this was tested with a single analyte under investigation. The scale up to 5 or 6 analytes within one sample will make the picture more complicated. For 5 analytes 10 different antibodies have to be used (5 analyte specific, 5 auxiliary). This will inevitably complicate the system and require thorough optimisation.

The data presented here were obtained within less than one year. We hope to come close to a workable system within the entire project duration of 3 years.

3.2.5 Acknowledgement

The RIANA project is funded under the Environment And Climate Programme 1994 - 1998 / (Area 2) Environmental Technologies (Instruments, Techniques And Methods For Monitoring The Environment) ENV4-CT95-0066

3.2.6 References

Abuknesha, R., Brecht, A. (1997): Multi-analyte immunoanalysis: Practical aspects of manufacturing transducers. Biosens. Bioelectron. 3, 159-160.

Blanchard, A.P., Kaiser, R.J., Hood, L.E. (1996): High-density oligonucleotide arrays. Biosens. Bioelectron. 11(6-7), 687-690.

Brecht, A., Abuknesha, R. (1995): Multi-analyte immunoassays application to environmental analysis. Trends Anal. Chem. 14(7), 361-371.

Chiron, S., Barceló, D. (1993): Determination of pesticides in drinking water by on-line solid-phase disk extraction followed by various liquid chromatographic systems. J. Chromatogr. 645(1), 125-133.

Harris, R.D., Wilkinson, J.S. (1995): Waveguide surface plasmon resonance sensors. Sensors and Actuators B 29, 261-267.

Marco, M.-P., Chiron, S., Gascón, J., Hammock, B.D., Barceló, D. (1995): Validation of two immunoassay methods for environmental monitoring of carbaryl and 1-naphthol in ground water samples. Anal. Chim. Acta 311, 319-329.

Meulenberg, E.P., Stoks, P.G. (1995): Water quality control in the production of drinking water from river water. Anal. Chim. Acta 311, 407-413.

Piehler, J., Brecht, A., Geckeler, K.E., Gauglitz, G. (1996): Surface modification for direct immunoprobes. Biosens. Bioelectron. 11(6-7), 579-590.

Tijssen, P. (1985): Practice and theory of enzyme immunoassays, Elsevier, Amsterdam.

Van Emon, J.M., Gerlach, C.L. (1995): The right environment for the immunoassay. Chemtech 11, 51-54.

Wadkins, R.M., Golden, J.P., Ligler, F.S. (1995): Calibration of biosensor response using simultaneous evanescent wave excitation of cyanine-labelled capture antibodies and antigens. Anal. Biochem. 232(1), 73-78.

3.3 Strategies for Recombinant Antibody Library Synthesis: An Advanced Source for Immunoglobulins in Environmental Analysis

Karl Kramer[1] **and Achim Knappik**[2]

[1]Department of Botany, Technical University of München at Freising-Weihenstephan, Alte Akademie 12, D-85350 Freising, Germany

[2]MorphoSys GmbH, Am Klopferspitz 19, D-82152 München, Germany

Abstract. Immunochemical analysis for environmental monitoring is so far predominantly based on antibodies derived from sera and hybridoma cultures. This conventional technology will be supported or even replaced in the near future by the recombinant antibody (Rab) approach, since Rab are accessible to precise modification of antibody properties by virtue of genetic engineering. Furthermore, the *in vitro* generation of Rab libraries bearing an antibody diversity comparable to the vertebrate immune repertoire paves the way to substitute animals as the antibody source of choice. Utilising the *s*-triazine herbicides as target molecules, various strategies for the synthesis of Rab libraries are presented comprising natural, semisynthetic and synthetic antibody genes. The Rab genes are expressed as fusion proteins at the surface of phage particles. Special attention is given to the establishment of an efficient selection procedure. Phages presenting specific Rab are isolated via immuno affinity chromatography employing *s*-triazine coated beads. The described libraries are designed as a gene pool to provide an array of antibody variants for future immunosensor applications in environmental analysis.

3.3.1   Introduction

The biological recognition unit constitutes an essential component of biosensor designs, since it hinges directly on the analytical quality of the particular sensor. A significant portion of sensor applications is based on antibody molecules (e.g. Minunni et al. 1995) among others, e.g. enzymes (Ayyagari et al. 1995), microbes (Rainina et al. 1996), receptors (Cheskis and Freedman 1996), nucleic acids (Watts et al. 1995), etc. This is mainly due to the capability of antibodies to specifically recognise a broad variety of substances and to bind the target molecules at an appropriate affinity level.

Antibodies are commonly obtained by immunising vertebrates and either collecting the serum which contains polyclonal antibodies or fusing antibody secreting B-lymphocytes with myeloma cells in order to produce monoclonal antibodies. However, during the last few years advances in genetic engineering facilitated the production of antibody fragments in bacteria thus avoiding the animal passage (Skerra and Plückthun 1988, Better et al., 1988). The subsequent combination of the recombinant antibodyidea with the phage display strategy (Smith 1985, McCafferty et al., 1990, Kang et al. 1991) finally established the basic instruments for the generation and convenient handling of antibody libraries comprising a portion or the complete variety of a vertebrate immune repertoire. The linkage of the antibody genotype and phenotype which constitutes a fundamental prerequisite for the clonal selection and affinity maturation of antibody secreting B-lymphocytes *in vivo* could thereby be mimicked *in vitro* with phages presenting the enclosed antibody gene as a functional protein at their surface.

3.3.2   Principle of Rab libraries utilising the phage display strategy

Following mRNA isolation (preferably from hybridoma cells, B-lymphocytes or peripher blood lymphocytes) and first strand cDNA synthesis, the heavy and light chain variable region (V_H and V_L) genes (c.f. Figure 1) are selectively amplified by polymerase chain reaction (PCR) utilising species-specific primers. Since the V_H and V_L comprising heterodimers are lacking stabilising inter chain disulphide bonds as a native antibody, individual V-genes are randomly assembled via a peptide linker encoding DNA sequence by a subsequent PCR prior to cloning into the phagemid

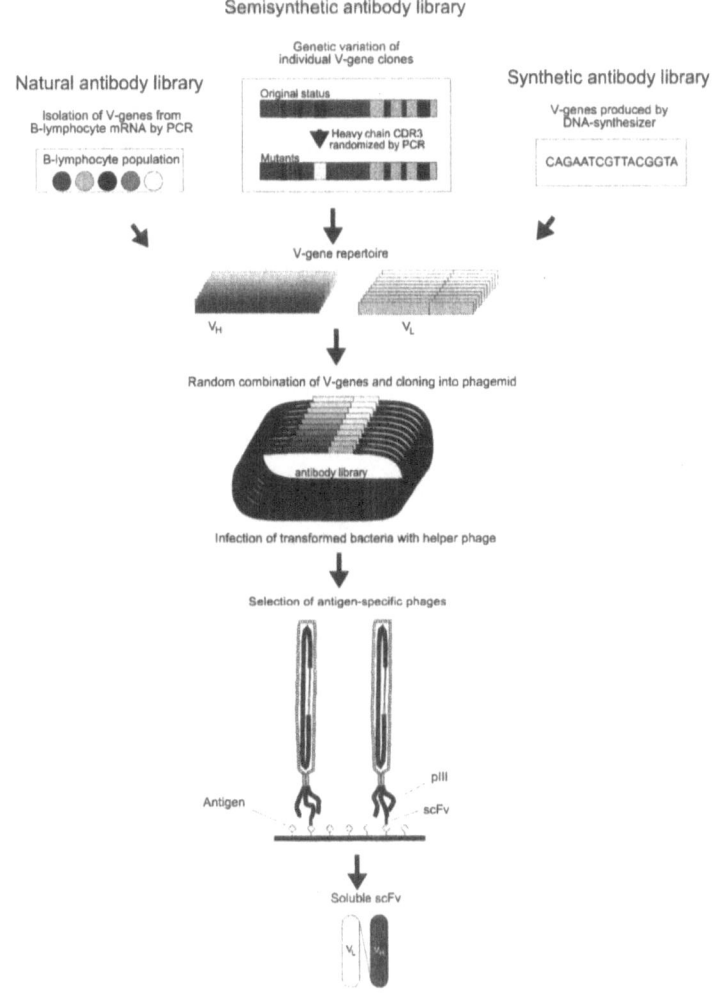

Figure 45: Principle of natural, semisynthetic and synthetic antibody libraries

vector. The assembled V-genes are commonly designated as single-chain F_v (scFv). The phagemid is drafted to juxtapose the scFv gene directly to the C-terminal moiety of the gene *III*, which encodes the minor phage coat protein III, in order to obtain a scFv-pIII fusion construct.

The scFv-gene *III* fusion construct is subsequently rescued by infecting phagemid-harbouring bacteria with helper phages, which provide all necessary genetic information for phage production. Initiation of expression results in the synthesis of complete phage particles, which are extruded from the bacterial periplasm into the culture supernatant. The phages contain a single-stranded phagemid encoding the scFv-pIII. The corresponding scFv protein is expressed and displayed at the phage surface with the pIII moiety providing a physical anchorage at the phage body. Specific phages are then selected at an antigen-coated solid phase, e.g. microtiter plates, petri dishes, cell culture flasks (Scott and Smith 1990, Marks et al. 1991), immunotubes (Hawkins et al. 1992), polystyrene or paramagnetic particles (Kramer and Hock 1995), or even on whole cells or tissue sections presenting the antigen at the surface. Specific phages are binding to the antigen whereas unbound phages are removed by repeated washes. Finally, the bound phages are eluted, and their genetic information is amplified by infecting suitable *E. coli* cells. The bacteria are plated in order to obtain single colonies for small scale expression of recombinant phage-associated or soluble scFv. Antigen-specificity of individual scFv-clones can be determined by ELISA on the basis of scFv-containing culture supernatant. Further characterisation of particular scFv is performed after DNA sequencing and protein purification.

3.3.3   Semisynthetic antibody libraries

The feasibility of modifying the analytical properties of an already existing antibody is claimed to constitute one of the essential advantages of the Rab strategy compared to classical antibody production schemes, e.g. antibodies derived from sera or hybridoma cultures. The modifications are achieved by artificially introducing sequence variations into a given antibody molecule (c.f. Figure 45). The resulting molecule is a semisynthetic hybrid consisting of synthetic sequences embedded into a natural antibody backbone structure. This approach represents the dominating type of semisynthetic library concepts along with the random pairing of heavy and light chain genes.

The refinement of a particular Rab structure by semisynthetic strategies is currently pursued on the basis of the *s*-triazine specific Rab K47H. This Rab was derived by cloning the V_H and V_L encoding genes of the corresponding *s*-triazine specific hybridoma line K4E7 (Kramer and Hock 1996). The cross-reactivity patterns of the obtained scFv and the parental monoclonal antibody are depicted in Figure 46. Libraries were generated by means of genetic engineering to create specificity and affinity variants.

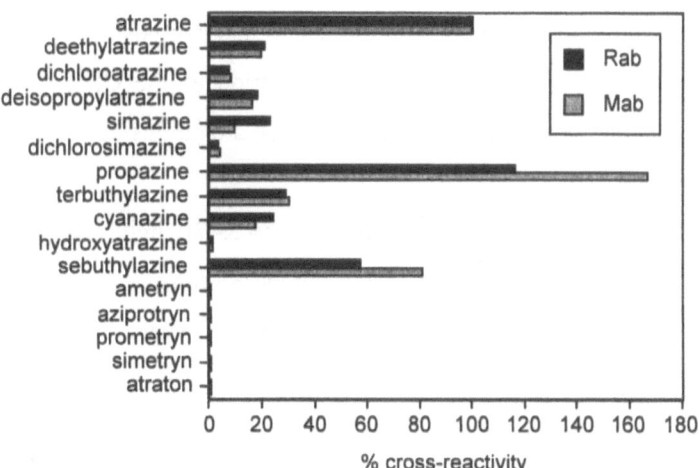

Figure 46: Cross-reactivity pattern of the recombinant scFv K47H (Rab) and the corresponding monoclonal antibody K4E7 (Mab) to 15 *s*-triazines related to atrazine

Two basically different concepts are applied in order to achieve a modification of the affinity and cross-reactivity pattern of K47H. The first technique designated as error prone PCR (Leung et al. 1989) introduces point mutations which are randomly distributed across the complete scFv gene. The *in vivo* occurring process of antibody maturation as a result of somatic hypermutation is herewith mimicked by virtue of a PCR reaction characterised by low sequence fidelity. Although the variation is restricted between three to five amino acid substitutions per mutated Rab molecule, a change in specificity and a 30-fold increase in affinity has been already described for the error prone strategy (Gram et al. 1992).

A second strategy applied for the variation of the original K47H scFv nucleotide sequence is entitled CDR walking (Barbas et al. 1994). In contrast to random point mutagenesis of the entire scFv moiety by error prone PCR, CDR walking is restricted to defined scFv sections predominantly involved in the hapten-antibody interaction. These sections are composed of six loops or complementarity-determining regions (CDRs), half of them derived from the heavy and the light chain, respectively. Variations between individual antibodies in the length and composition of the CDRs contribute to the plasticity of the binding site. Variation of the CDR sequence is therefore regarded as the most efficient mean to initiate alterations of the affinity and cross-reactivity pattern.

Figure 47 exhibits the nucleotide sequence of the K47H light and heavy chain variable regions. The CDRs are underlined and numbered consecutively as CDRH1, CDRH2 and CDRH3 for the heavy chain moiety and CDRL1, CDRL2 and CDRL3 for the light chain moiety, respectively. Essential contact sites to *s*-triazines are labelled by asterisks. They are thought to be located in the CDRH1, CDRH3 and CDRL3 based on molecular modelling (Hörsch et al. 1996). In order to introduce a random sequence of 18 nucleotides into the CDRH3 (Figure 47), a technique known as splicing by overlap extension was employed (Barbas et al. 1994). The scFv K47H was divided into two sections at the 5' end of CDRH3. Both scFv moieties were amplified in separate PCRs using an primer which randomised the HCDR3 region. The scFv moieties were subsequently fused again in a subsequent PCR to obtain the complete, CDRH3-randomized scFv gene.

K47H V$_H$ domain

CAGGTGAAACTGCAGCAGTCAGGAGGAGGCTTGGTGCAACCTGGAGGATCCATGAAACTCTCTTGTGCTGCCTC

TGGATTCACTTTCAGTGAC*GTCTGGATGGAC*TGGGTCCGCCAGTCTCCAGAGAAGGGGCTTGAGTGGGTTGCTG
            ***   ***

AAATTAGAAACAAAGCTAATAATCATGCAGCATATTATGCTGAGTCTGTGAAAGGGGAGGTTCACCGTTTCAAGA

GATTCCAAAAGTAATGTCTACCTGCACATGAATAGCTTAAGACCTGAAGACACTGGCATTTATTATTGTACCAG

GATG*CACAGCTATAGGTACGA*CGGGTTTGCTTACTGGGGCCAAGGGACCACGGTCACCGTCTCC
     ***   ******   ***

K47H V$_L$ domain

GACATCGAGCTCACTCAGTCTCCAGCTTCTTTGGCTGTGTCTCTAGGGCAGAGGGCCACCATCTCCTGCAGAGC

CAGTGAAAGTGTTGATATTTATGGCAATAGTTTTATGCACTGGTACCAGCAGAAACCAGGACAGCCACCCAAAC

TCCTCATCTATCGTGCATCCAACCTAGAAACTGGGATCCCTGCCAGGTTCAGTGGCAGTGGGTCTAGGACAGAC

TTCACCCTCACCATTAATCCTGTGGAGGCTGATGATGTTGCAACCTATTTCTGT*CAGCAAAGTAAATCAGCTCC*
                          ***   ******   ***

*GTACA*CGTTCGGAGGGGGCACCAAGCTGGAAATCAAACGGGCG
 ***

Figure 47: CDR walking strategy based on the K47H nucleotide sequence. The three regions encoding the CDRs of the individual domains are underlined. Nucleotide sections of CDRH1, CDRH3 and CDRL3 are subjected to randomisation as indicated by italics. Supposed contact sites to atrazine are labelled by asterisks

The CDRH3-randomized libraries will be selected for clones exhibiting increased binding affinity to the desired *s*-triazine metabolites. Selected clones can be subjected to a second cycle of randomisation targeting the CDRL3 (Figure 46) while maintaining the residual antibody structure. The stringency of the selection conditions will be increased in order to enrich sequence variants in the CDRL3 library with improved binding behaviour to the desired target molecule. Finally, the CDRH1 will be subjected to randomisation. Thereby CDRs which contact the ligand will be stepwise modified. Every randomisation is combined with an increased level of selection stringency in order to optimise the antibody- antigen interaction by the evolutive concept of selection pressure and individual fitness. This CDR walking strategy has been successfully applied by Yang et al. (1995) varying the CDRs sequentially (as outlined above) or in parallel yielding up to a 420-fold increase in affinity for an anti-HIV antibody.

3.3.4 *s*-Triazine specific library based on natural antibody genes

Whereas the above mentioned Rab libraries are directed to the specificity and affinity refinement of a given Rab sequence, alternative strategies are focusing on the generation of appropriate antibody gene pools, which can be used to directly select antibodies with desired affinity and specificity. Moreover, individual variants can be further optimised, if necessary, by the methods described above. A comprehensive, *s*-triazine specific library is presently built on the basis of lymphocyte populations previously stimulated by *s*-triazines coupled to immunogenic carrier proteins. Eighteen out of 87 different *s*-triazine stimulated lymphocyte populations were chosen for library construction, since titer validation by ELISA revealed specific recognition of chlorinated *s*-triazines containing a dealkyl-, *tert.* buthyl-, ethyl-, and isopropyl-group, respectively, for these 18 populations. Lymphocytes bearing *s*-triazine specific antibody genes are currently selected by virtue of immunomagnetic separation (IMS). IMS is based on the fact that hybridoma cells express surface bound antibody molecules. Cells expressing the desired antibody specificity can be tagged by paramagnetic particles (Advanced Magnetics, MA, USA) previously coated with the corresponding triazine-derivatives (Kramer et al. 1994). Up to 2×10^5 cells (average 1×10^4) out of a total of 5×10^7 B-lymphocytes were obtained per IMS in the magnet-bound fraction. These positively selected B-lymphocytes will comprise a library of approximately 10^6 independent clones upon completion, which is highly enriched for triazine binders by the IMS method.

3.3.5   Synthetic antibody library

A novel approach towards high-affinity antibodies with tailored specificity without the need of immunising animals is based on a set of 49 fully synthetic human antibody sequences (Knappik et al. in preparation). Although this technology was mainly developed to create human antibodies with low immunogenicity useful in immunotherapy, it can also be used to create specific binders for applications in the diagnostic and reagent areas. The synthetic chains were derived from an comprehensive analysis of the human antibody repertoire, which is now known to consist of about 50 V_H and 70 V_L germline genes (Tomlinson et al. 1992, Cox et al. 1994, Williams et al. 1996). Comparison of rearranged human antibody sequences with this germline repertoire showed that many of the germline genes are never used during an immune response, reducing the actual used germline repertoire to 25 V_H and 27 V_L sequences. These sequences could be grouped into 7 V_H and 7 V_L subfamilies, respectively, based on both sequence homology and structural diversity (Chothia and Lesk 1987, Chothia et al. 1989, Chothia et al. 1992). For each subfamily a consensus gene was deduced and chemically synthesised, leading to 7 x 7 different V_H-V_L combinations which formed the basis for the synthetic Human Combinatorial Antibody Library (HuCAL).

An initial diversity of more than $2x10^9$ independent clones was introduced by replacing both the V_L and V_H CDR3s with randomised oligonucleotide cassettes created by TRIM technology (Virnekäs et al. 1994). These TRIM cassettes were designed by analysing the CDR3 loops of existing human antibodies and hence mimic the composition and length distribution of natural antibodies. Since the V_H and V_L genes were assembled by chemically synthesised oligonucleotides, unique restriction sites could be designed and introduced at each end of all six CDRs. This modular gene structure makes it now for the first time possible to rapidly access each of the six CDRs of binders obtained from the initial library, which allowes affinity and specificity optimisation by a CDR walking procedure (Yang et al. 1995, Barbas and Burton 1996) using pre-built oligonucleotide libraries.

A subset of the initial, CDR3-randomized library (1 out of the 49 HuCAL genes was used) has been selected for variants binding to a simazine derivative. Following two cycles of selection by immunoaffinity chromatography (c.f. below), an aliquot of 96 independent clones was expressed on small scale and tested by phage ELISA. Phage containing culture supernatant was applied onto microtiter wells precoated with the simazine derivative. After staining bound phages by HRP-conjugated anti-phage antibodies, 38 out of 96 clones evoked specific signals in the substrate reaction, whereby a minimum of 100 % increase in absorbance signal (corresponding to $OD_{450\ nm}$ of 0.1 - 0.3, compared to $OD_{450\ nm} = 0.03$ in uncoated control wells) was demanded for revealing binding specificity.

The future success of recombinant antibody fragments derived from phage display libraries will largely depend on high expression yields, since they are needed in large amounts for use as therapeutic agents, diagnostic reagents and for biochemical research (Winter et al. 1994). However, very often problems in the production of antibody fragments in *E. coli* have been observed (Somerville et al. 1994, Sawyer et al. 1994, Ayala et al. 1995). Expression yields were shown to vary widely and so far only very few publications reported expression yields in the range of several mg of functional, soluble protein per litre of shaking flask culture. It is known that expression yields in *E. coli* depend for example on the codon usage of a particular gene (Hernan et al. 1992). Additionally it was shown that the antibody amino acid sequence itself determines efficient folding and therefore yield (Duenas et al. 1995, Knappik and Plückthun 1995). The synthetic approach described here created the possibility to overcome expression problems already at the beginning of the selection process by optimising the codon usage of each of the 49 HuCAL genes and by removing amino acid residues which introduce a bottleneck in the folding pathway. Indeed it was found that HuCAL antibodies are produced as functional molecules in the multi-milligram range per litre *E. coli* culture (Knappik et al. unpublished results).

3.3.6   Selection of Rab libraries

The difficulties encountered by specific selection of desired clones is regarded as a serious bottleneck of the phage display strategy. We therefore examined different facilities for the reliable and precise isolation of appropriate clones. Phages displaying hapten-specific scFv were separated in model experiments from unspecific phages by adsorption of an atrazine derivative via carrier protein onto microtiter plates for panning, by immobilisation at the surface of immunomagnetic particles for immunomagnetic selection (IMS), and by direct coupling on sepharose for immunoaffinity chromatography in columns (IAC). Specific phages were enriched by subjecting the libraries to repetitive rounds of selection, with each consisting of binding, washing and elution steps, reinfection and expansion of bacteria to re-amplify the specifically bound phage fraction.

The focus on the selection systems IMS and IAC was due to the fact that with this systems hapten immobilisation does not require a selection surface characterised by high protein binding capacity as for the classical panning procedure in microtiter plates. Panning frequently encounters high backgrounds in the selection step since phage bodies tend to bind unspecifically at these panning devices. Apart from this IAC is superior compared to panning and IMS, since carboxylated *s*-triazine derivatives can be directly coupled to sepharose avoiding the carrier protein conjugation. Thereby, undesired carrier protein-specific Rab variants included particularly in highly diverse libraries are not enriched. Furthermore, IAC columns once prepared are re-usable and the selection process can be conveniently automated by virtue of peristaltic pumps.

Table 20: IAC-Separation of scFv-randomized library with atrazine-derivative (Ipr/Cl/C5). The error prone-randomized library derived from the K4H7-scFv was utilized for phage IAC selection. atrazine-coupled sepharose columns were incubated with the corresponding number of phages which was determined as colony forming units (total cfu). After washing with 80 ml wash buffer, bound phages were eluted with 20 ml glycine-HCl, precipitated und employed for reinfection with TG1 bacteria. Bacteria were plated in serial dilutions onto SOB plates containing 100 µg ampicillin and 2% (w/v) glucose and incubated overnight at 37 °C. Phage-bearing bacteria were counted as colony forming units (separated cfu)

Separation cycle	Separation conditions	Total cfu	Separated cfu	Enrichment factor
1	30 min/ 80 ml	9.75×10^8	1.13×10^6	863
2	15 min/ 80 ml	6.60×10^8	9.30×10^5	710
3	1 min/ 80 ml	2.82×10^9	3.57×10^6	789

The K47H-based library which was generated by error prone PCR as described above was subjected to repetitive cycles of separation applying the IAC concept (Kretzschmar et al. 1995). Details of the selection process are summarized in Table 20. The initial library was selected for variants with improved affinities to an atrazine derivative by increasing the stringency (c.f. separation conditions) of the consecutive selection steps. At each separation cycle aliquots of the separated fraction were tested by ELISA for atrazine binding (c.f. Figure 48). Clones of the individual fractions were grouped according to their binding behaviour into five different categories: no change of binding behaviour compared to the unmutated triazine-specific scFv, decreased or increased affinity, complete loss of binding capability and finally, atrazine binding clones which are not displacable by the analyte. The percentage of the first two categories were diminished with increased separation stringency. This may be due to the limited affinity of antibody variants which did not match the stringent conditions applied in the higher selection steps. In contrast, a significant amount of approximately 20 % of the total antibody clones was detected even after the third separation cycle exhibiting an increased affinity.

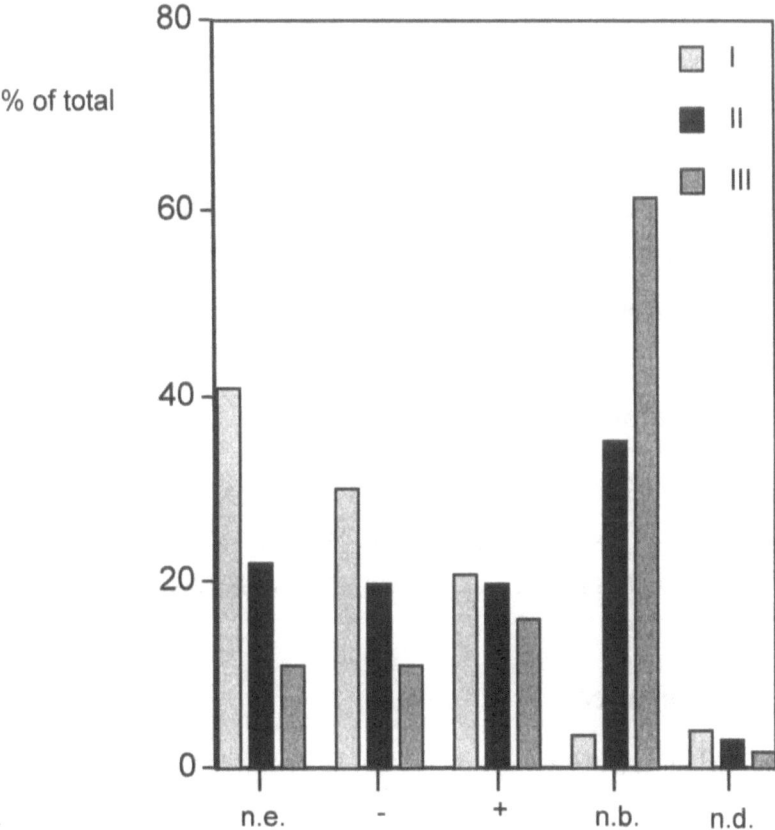

Figure 48: Distribution of binding variants in the separated fraction of different separation cycles (I, II, III). Aliquots of 180-190 clones were tested by ELISA for atrazine-binding. Results were grouped into following categories: no effect on binding behaviour (n.e.); decreased affinity (-); increased affinity (+); complete loss of binding capability (n.b.); atrazine binding clones which are not displacable by the analyte (n.d.). The results are based on the calibration curve of the unmutated atrazine-specific scFv K47H

3.3.7 Perspectives

A few groups are meanwhile involved in Rab synthesis for environmental analysis (e.g. Ward et al. 1993, Karu et al. 1994, Graham et al. 1995, Kamps-Holtzapple and Stanker 1996), indicating that Rab will become a valuable tool for the monitoring of xenobiotics in the next future. As the cloning of already existing monoclonal antibodies was commonly regarded as an initial effort to enter this promising technique, the generation of comprehensive antibody libraries constitutes the consequent step beyond to match the demands of environmental analysists. Expansion and diversification of the libraries followed by exchanging these antibody pools between different research groups will potentially accelerate the development of appropriate binding molecules for immunosensor applications.

3.3.8 Acknowledgements

We thank Prof. B. Hock for his helpful suggestions on the manuscript and Mrs. Vater and Mrs. Reichle for their technical assistance at the library construction. Part of this work was supported by the European Community (ENV4-CT96-0333).

3.3.9 References

Ayala, M., Balint, R.F., Fernandez de Cossio, L., Canaan Haden, J.W., Larrick, J.W., Gavilondo, J.V. (1995): Variable region sequence modulates periplasmic export of a single-chain Fv antibody fragment in *Escherichia coli*. Biotechniques 18, 832, 835-8, 840-2.

Ayyagari, M.S., Pande, R., Kamtekar, S., Gao, H., Marx, K.A., Kumar, J., Tripathy, S.K., Akkara, J.A., Kaplan, D.L. (1995): Molecular assembly of proteins and conjugated polymers: Toward development of biosensors. Biotechnology & Bioengineering 45, 116 - 121.

Barbas, C.F., Burton, D.R. (1996): Selection and evolution of high-affinity human anti-viral antibodies. TibTech 14, 230-234.

Barbas, C.F., Hu, D., Dunlop, N., Sawyer, L., Cababa, D., Hendry, R.M., Nara, P.L., Burton, D.R. (1994): In vitro evolution of a neutralizing human antibody to human immunodefiency virus type 1 to enhance affinity and broaden strain cross-reactivity. Proc. Natl. Acad. Sci. USA 91, 3809-3813.

Better, M., Chang, C.P., Robinson, R.R., Horwitz, A.H. (1988): *Escherichia coli* secretion of an active chimeric antibody fragment. Science 240, 1041-1043.

Cheskis, B., Freedman, L.P. (1996): Modulation of nuclear receptor interactions by ligands: Kinetic analysis using surface plasmon resonance. Biochemistry 35, 3309 - 3318.

Chothia, C., Lesk, A.M. (1987): Canonical structures for the hypervariable regions of immunoglobulins. J. Mol. Biol. 196, 901-917.

Chothia, C., Lesk, A.M., Gherardi, E., Tomlinson, I.A., Walter, G., Marks, J.D., Llewelyn, M.B., Winter, G. (1992): Structural repertoire of the human VH segments. J. Mol. Biol. 227, 799-817.

Chothia, C., Lesk, A.M., Tramontano, A., Levitt, M., Smith-Gill, S.J., Air, G., Sheriff, S., Padlan, E.A., Davies, D., Tulip, W.R., Colman, P.M., Spinelli, S., Alzari, P.M., Poljak, R.J. (1989): Conformations of immunoglobulin hypervariable regions. Nature 342, 877-883.

Duenas, M., Ayala, M., Vazquez, J., Ohlin, M., Soderlind, E., Borrebaeck, C.A.K., Gavilondo, J.V. (1995): A point mutation in a murine immunoglobulin V-region strongly influences the antibody yield in escherichia coli. Gene 158, 61-66.

Cox, J.P., Tomlinson, I.M., Winter, G. (1994): A directory of human germ-line V kappa segments reveals a strong bias in their usage. Eur. J. Immunol. 24, 827-836.

Graham, B.M., Porter, A.J.R., Harris, W.J. (1995): Cloning, expression and characterisation of a single-chain antibody fragment to the herbicide paraquate. J. Chem. Technol. Biotechnol. 63, 279-289.

Gram, H., Marconi, L.-A., Barbas, C.F., Collet, T.A., Lerner, R.A., Kang, A.S. (1992): *In vitro* selection and affinity maturation of antibodies from naive combinatorial immunoglobulin library. Proc. Natl. Acad. Sci. USA 89, 3576-3580.

Hawkins, R.E., Russell, S.J., Winter, G. (1992): Selection of phage antibodies by binding affinity: Mimicking affinity maturation. J. Mol. Biol. 226, 889-896.

Hernan, R.A., Hui, H.L., Andracki, M.E., Noble, R.W., Sligar, S.G., Walder, J.A., Walder, R.Y. (1992): Human hemoglobin expression in *Escherichia coli*: Importance of optimal codon usage. Biochemistry 31, 8619-8628.

Hörsch, S., Pleiss, J., Kramer, K., Schmid, R.D. (1996): K411B, a triazine-binding single-chain antibody: Structure modelling and hapten docking. 10th Spring-Workshop "Molecular Modelling", Darmstadt, Germany.

Kamps-Holtzapple, C., Stanker, L.H. (1996): Development of recombinant single-chain variable portion recognizing potato glycoalkaloids, p. 485-499. In: Immunoassays for residue analysis (Beier, R.C., Stanker, L.H., eds.). ACS Symposium Series 621.

Kang, A.S., Barbas, C.F., Janda, K.D., Benkovic, S.J., Lerner, R.A. (1991): Linkage of recognition and replication functions by assembling combinatorial Fab libraries along phage surfaces. Proc. Natl. Acad. Sci. USA 88, 4363.

Karu, A.E., Scholthof, K.-B.G., Zhang, G., Bell, C.W. (1994): Recombinant antibodies to small analytes and prospects for deriving them from synthetic combinatorial libraries. Food Agricult. Immunol. 6, 277-286.

Knappik, A., Plückthun, A. (1995): Engineered turns of a recombinant antibody improve its in vivo folding. Protein Eng. 8, 81-89.

Kramer, K., Giersch, T., Hock, B. (1994): Magnetic bead selection of hybridomas producing pesticide antibodies. Food Agricult. Immunol. 6, 5-16.

Kramer, K., Hock, B. (1995): Rekombinante Antikörper in der Umweltanalytik. Lebensmittel- & Biotechnologie $\underline{2}$, 49-56.

Kramer, K., Hock, B. (1996): Recombinant single-chain antibodies against s-triazines. Food Agricult. Immunol. $\underline{8}$, 97-109.

Kretzschmar, T., Zimmermann, C., Geiser, M. (1995): Selection procedures for nonmatured phage antibodies: A quantitative comparison and optimization strategies. Anal. Biochem. $\underline{224}$, 413-419.

Leung, D.W., Chen, E., Goeddel, D.V. (1989): A method for random mutagenesis of a defined DNA segment using a modified polymerase chain reaction. J. Meth. Cell. Mol. Biol. $\underline{1}$, 11-15.

Marks, J.D., Hoogenboom, H.R., Bonnert, T.P., McCafferty, J., Griffiths, A.D., Winter, G. (1991): By-passing immunization: Human antibodies from V-gene libraries displayed on phage. J. Mol. Biol. $\underline{222}$, 581-597.

McCafferty, J., Griffiths, A.D., Winter, G., Chiswell, D.J. (1990): Phage antibodies: Filamentous phage displaying antibody variable domains. Nature $\underline{348}$, 552.

Minunni, M., Mascini, M., Guilbault, G.G., Hock, B. (1995): The quartz crystal microbalance as biosensor. Anal. Letts. $\underline{28}$, 749-764.

Rainina, E.I., Badalian, I.E., Ignatov, O.V., Fedorov, A.Y., Simonian, A.L., Varfolomeyev, S.D. (1996): Cell biosensor for detection of phenol in aqueous solutions. Appl. Biochem. Biotechnol. $\underline{56}$, 117-127.

Sawyer, J.R., Schlom, J., Kashmiri, S.V. (1994): The effects of induction conditions on production of a soluble anti-tumor sFv in *Escherichia coli*. Protein Eng. $\underline{7}$, 1401-1406.

Scott, J.K., Smith, G.P. (1990): Searching for peptide ligands with an epitope library . Science $\underline{249}$, 386-390.

Skerra, A., Plückthun, A. (1988): Assembly of a functional immunoglobulin Fv fragment in *Escherichia coli.* Science $\underline{240}$, 1038-1041.

Smith, G.P. (1985): Filamentous fusion phage: Novel expression vectors that display cloned antigens on the virion surface. Science 228, 1315–1316.

Somerville, J.E., Goshorn, S.C., Fell, H.P., Darveau, R.P. (1994): Bacterial aspects associated with the expression of a single-chain antibody fragment in *Escherichia coli*. Appl. Microbiol. Biotechnol. 42, 595-603.

Tomlinson, I.M., Walter, G., Marks, J.D., Llewelyn, M.B., Winter, G. (1992): The repertoire of human germline VH sequences reveals about fifity groups of VH segments with different hypervariable loops. J. Mol. Biol. 227, 776-798.

Virnekäs, B., Ge, L., Plückthun, A., Schneider, K.C., Wellnhofer, G., Moroney, S.E. (1994): Trinucleotide phosphoramidites: Ideal reagents for the synthesis of mixed oligonucleotides for random mutagenesis. Nucl. Acid. Res. 22, 5600-5607.

Ward, V.K., Schneider, P.G., Kreißig, S.B., Hammock, B.D., Choudary, P.V. (1993): Cloning, Sequencing and expression of the Fab fragment of a monoclonal antibody to the herbicide atrazine. Prot. Eng. 6, 981-988.

Watts, H.J., Yeung, D., Parkes, H. (1995): Real-time detection and quantification of DNA hybridzation by an optical biosensor. Anal. Chem. 67, 4283-4289.

Williams, S.C., Frippiat, J.P., Tomlinson, I.M., Ignatovich, O., Lefranc, M.P., Winter, G. (1996): Sequence and evolution of the human germline V-lambda repertoire. J. Mol. Biol. 264, 220-232.

Winter, G., Griffiths, A.D., Hawkins, R.E., Hoogenboom, H.R. (1994): Making antibodies by phage display technology. Annu. Rev. Immunol. 12, 433-455.

Yang, W.P., Green, K., Pinzsweeney, S., Briones, A.T., Burton, D.R., Barbas, C.F. (1995): CDR walking mutagenesis for the affinity maturation of a potent human anti-HIV-1 antibody into the picomolar range. J. Mol. Biol. 254:392-403.

3.4 Characterization of a Monoclonal Antibody and its Fab Fragment Against Diphenylurea Hapten with BIA

Walter F.M. Stöcklein[*], Axel Warsinke, Burkhard Micheel, Wolfgang Höhne[1], Gerhard Kempter[2] and Frieder W. Scheller

Institute of Biochemistry and Molecular Physiology, University of Potsdam, c/o Max-Delbrück Centre, Robert-Roessle-Str. 10, 13122 Berlin, Germany

[1]Institute of Organic Chemistry and Structure Analytics, University of Potsdam, [2]Institute of Biochemistry, Humboldt University, Berlin, Germany

Abstract. Biospecific interaction analysis was used for the detection of two diphenylurea derivatives and the characterization of a monoclonal antibody (Mab) and the derived Fab fragment. The k_{on} and k_{off} rates were determined. Affinity constants were obtained from kinetic data and from affinity in solution for the binding of antigen I and the modified antigen II containing the coupling group. The binding of Fab to immobilized antigen was enthalpy driven below 37°C. The affinities were decreased in the presence of ethanol, and the crossreactivity was shifted in favour of antigen II. At pH values below 4.5 only antigen II (lacking the aromatic primary amino group) was bound by Fab. The detection limit for antigen I was governed by the K_D, for antigen II by the measuring system. 50% inhibition was achieved with 0.6 ppb antigen II. The assay time was 15 min.

---

[*] to whom correspondence should be addressed

3.4.1   Introduction

One of the main groups of photosystem II inhibiting herbicides is structurally related to phenylurea. Isoproturon and Chlortoluron are among the 12 most frequently detected pesticides in water samples in Germany (Wolter 1995). The European limit for individual pesticides in drinking water (the EU Maximum Admissible Concentration, MAC) is 0.1 µg/l.

The methods primarily used for the determination of phenylurea derivatives are GC (after derivatization) or HPLC with UV detection. Several enzyme linked immunoassays have been developed for the measurement of phenylurea pesticides, e.g. isoproturon (Katmeh et al. 1994) or chlortoluron (Katmeh et al. 1996), even at concentrations below the MAC limit. Whole-cell biosensors have been described in literature, but their sensitivity and selectivity are poor compared with enzyme- or antibody-based methods, as reviewed by Krämer and Schmid (1992).

The preparation and characterization of antibodies (or fragments) with high affinity and appropriate specificity is a prerequisite for the development of immunoassays. Various methods can be used for kinetic analyses, e.g. fluorescence titration, but optical systems based on surface plasmon resonance can be regarded to be the most versatile and accurate instruments for biospecific interaction analysis (Neri et al. 1996).

The results presented here were obtained with a commercial surface plasmon resonance based system for biospecific interaction analysis, BIACORE® 2000. Monoclonal antibodies and Fab fragments against N-(2-N-chloroacetyl-aminobenzyl)-N′-4-chlorophenylurea (antigen II) were prepared. The on- and off- rates were determined by direct interaction analysis with immobilized antigen under various conditions. Affinity constants were obtained from the ratio of on- and off-rates or by measurements of affinity in solution. The influence of the antigen coupling group, immobilization, temperature, ethanol and pH on the Fab - antigen binding was investigated.

3.4.2   Materials and methods

3.4.2.1   Equipment and reagents

The BIACORE™ 2000 system, sensor chips CM5, buffer HBS (10 mM Hepes with 0.15 M NaCl, 3.4 mM EDTA and 0.005% surfactant P20 at pH 7.4), amine coupling kit, 2-(2-pyridinyldithio) ethaneamine (PDEA), and BIA evaluation software were obtained from Biacore AB, Uppsala, Sweden. Cystamine was purchased from Fluka, Deisenhofen, Germany, dithiothreitol and dimethylsulfoxide from Merck, Darmstadt, Germany. The phenylurea derivatives were chemically synthesized in our group: N-(2-aminobenzyl)-N′-4-chloro-phenylurea (antigen I) and N-(2-N-chloroacetyl-amino-benzyl)-N′-4-chlorophenyl-urea (antigen II).

Monoclonal antibody (Mab) B76-BF5 was prepared from mice. The immunogen was a antigen II - conjugate with 2-iminothiolate modified *Helix pomatia* hemocyanin (Lindner and Robey 1987). The Mab was purified by affinity chromatography using a Protein A sepharose column. Fab fragments were generated by papain cleavage of Mab and purified by gel filtration with Sephadex G-75, after the removal of the Fc fragment with protein A.

3.4.2.2   BIACORE experiments

All experiments were performed following the instructions of the Biacore manuals, unless otherwise stated. Standard conditions were: 25°C, flow rate 10 µl/min, running buffer HBS, pH 7.4.

Immobilization

The CM5 chips were activated using the surface thiol method with EDC/NHS and cystamine. The activated chip (containing SH groups after reduction with dithiothreitol) was reacted outside the instrument with 10 mg/ml antigen II in dimethylsulfoxide containing 1/100 vol. 0.1M NaHCO3 for 1 h. Excess reactive groups on the chip were blocked with PDEA and ethanolamine, and conditioned with three 10 µl injections of 30 mM HCl. The thioether linkage was stable under the experimental conditions, and the chip containing the ligand was used for several hundreds of binding experiments. The amount of immobilized antigen corresponded to a maximum increase of the SPR signal of about 0, 300, 1000 and 12000 resonance units (RU) after Mab injection to flowcell 1, 2, 3 and 4, respectively. Flowcell 1 was used as negative control, flowcells 2 and 3 for kinetics and flowcell 4 for binding experiments under mass transport limitation (affinity in solution and active antibody determination).

Binding kinetics

Various HCl solutions were tested for complete regeneration of the sensor chip. Complete regeneration was obtained by injection of 10 µl 50 mM HCl (pH 1.3).

The apparent association rate constant k_{on} was calculated from the slope of a plot of k_s vs. antibody concentration as described in the Biacore handbooks. 50 µl of antibody or Fab was injected (*kinject*) at concentrations between 10 and 200 nM. A linear dependence of k_s on C was found only for concentrations between 10 and 120 nM Fab and 5 to 50 nM antibody. A low ligand density was used to avoid mass transfer limitation (R_{max} for antibody saturation was about 300 RU).

The protein concentration for dissociation experiments was 2µM. In order to prevent rebinding, an excess of antigen II (3µM) was injected during the dissociation phase, using the coinject command. The active antibody concentration was found out using the method of mass transport limited binding (Karlsson et al. 1993).

Affinity in solution

Antibody or Fab (0.5 to 5 nM) was preincubated with antigen concentrations ranging from 0.01 to 1000 nM. After equilibration for at least 1 hour, the amount of free binding sites was measured by injecting 50 µl into a flowcell containing a high ligand density (flowcell 4). This procedure is based on a method described for enzyme linked immunoassays (Friguet et al. 1985). Linearity of the initial signal slope vs. antibody concentration was checked prior to these experiments. The relative initial signal slope was plotted against the logarithm of the antigen concentration. The resulting sigmoidal curve was fitted by a four-parameter log-logistic model (Microcal Origin, sigmoidal fit). The formula for the calculation of the binding constants K_D for Fab and bivalent Mab are described in the appendix (Karlsson 1994).

3.4.3 Results and discussion

3.4.3.1 Binding experiments

Kinetic constants can be obtained with BIACORE only if one of the binding partners is immobilized. With immobilized antibody, a conjugate of the antigen (MW about 300) with a protein should be used as the analyte, as the signal is too low for small molecules. Such conjugates are usually heterogeneous and do not allow quantitation of results. Therefore, the antigen was immobilized and Mab or Fab used as the analyte (Figure 49).

184

The binding of Mab to the immobilized ligand should be heterogeneous due to the avidity effect of bivalent binding. Currently, there exists no satisfactory evaluation method, which takes into account the avidity effect. The calculation of rate constants by applying the evaluation for 1:1 binding can lead to underestimated k_{off} values and overestimated k_{on} values. The results show that the rate constants for Mab indeed deviate from the constants for Fab as expected. Determination of k_{on} depends on which part of the association curve was selected for fitting. The apparent $K_D = k_{off} / k_{on}$ is 7.5 fold lower for Mab, compared with Fab (Table 21).

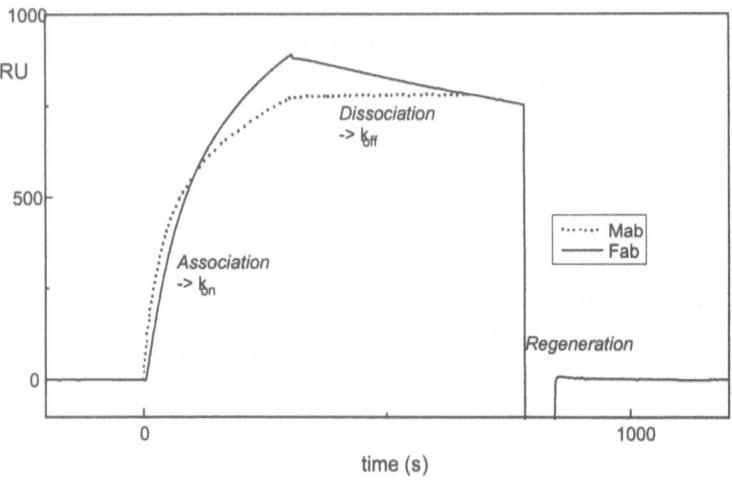

Figure 49: BIACORE Sensorgram for 100 nM Mab and 40 nM Fab

This deviation results from the slow dissociation of the antibodies, as bivalent binding is stronger than monovalent binding and rebinding counteracts dissociation. Addition of excess antigen during the Mab dissociation phase increased the dissociation rate, but not to the value obtained with the monovalent Fab.

However, sensorgrams obtained with Fab fragments had nearly ideal shape with respect to the fitting procedures used. Coinjection of high concentration of antigen during the dissociation did not affect the rate significantly at temperatures below 25°C.

A nearly twofold increase of k_{on} was only observed at 37°C using a high ligand density. Therefore the binding kinetics of the monoclonal antibody is best represented by the on- and off-rates obtained with Fab fragments.

Table 21: Binding data obtained with BIACORE® 2000 at 25°C

Protein	k_{on}	k_{off}	K_D (M)		
	$(M^{-1}s^{-1})$	(s^{-1})	II, imm.[a]	I, sol.[b]	II, sol.[b]
Mab	6.7E5	7.2E-4	1.1E-9	8E-9	3.4E-10
Fab	3.4E5	2.8E-3	8.2E-9	9E-9	2.3E-10

[a] $K_D = k_{off} / k_{on}$ for immobilized antigen II
[b] K_D from affinity in solution for antigen I and II

In contrast to the direct binding experiments, avidity should not affect the binding of antigen in solution. The only difference between Mab and Fab is that the true fraction of free antibody has to be calculated from the data (see section 3.4.2.2). Accordingly, the results of the ´affinity in solution´ experiments show that there are only minor differences in K_D obtained for Mab and Fab (Table 21, sol.).

It can be concluded from the binding data that the coupling group (chloroacetyl group) in antigen II is involved in high affinity binding, which increased the affinity by a factor of 26 (Fab) to 42 (Mab).

3.4.3.2   Temperature dependence of binding

Immunoassays are usually performed at temperatures between ´room temperature´ and 37°C (physiological temperature). It is commonly assumed that the affinity increases with temperature, as the activity of most enzymes does. However, antibody-antigen interactions for which hydrogen bonding is important are more stable at lower temperatures (´cold antibodies´, Tijssen 1985).

Table 22: Kinetic and thermodynamic data for Fab binding to immobilized antigen II

Temp. (°C)	5	15	25	35
K_A (x 10^{-8} M^{-1})	44.4	12.5	5.0	2.2
ΔG (kJ M^{-1})	-51.3	-50.1	-49.6	-49.2
ΔH (kJ M^{-1})	-88.9	-75.8	-64.0	-52.5
ΔS (J M^{-1} K^{-1})	-135.3	-89.2	-48.3	-10.7

Therefore, the affinity of Fab for immobilized antigen II was determined from on and off rates in the temperature range 5°C to 35°C (Table 22). The temperature dependence of affinity constants is shown by the van't Hoff plot, which was fitted by nonlinear regression (Figure 50). The values of ΔG, ΔH, ΔS and Δc_p were calculated as described in the appendix. Δc_p was 1.22 kJ mol^{-1} K^{-1}.

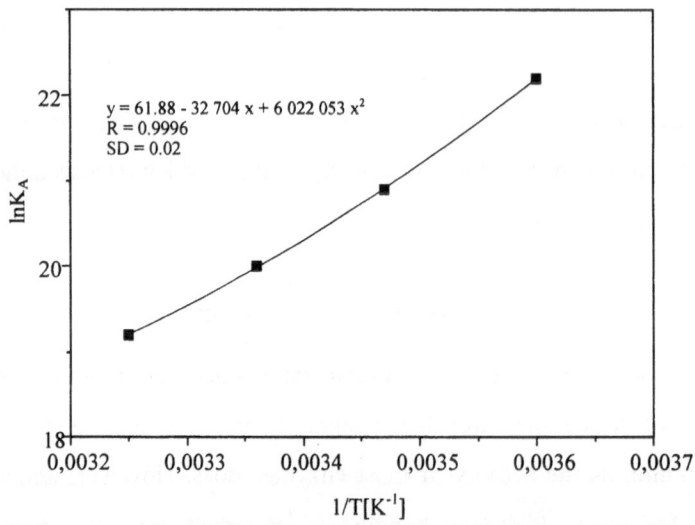

Figure 50: van't Hoff plot for the temperature dependent binding of Fab to immobilized antigen II

It is evident from the data that the increase of ΔH compensates largely the decrease of -T*ΔS. A similar temperature dependence and enthalpy-entropy compensation was described for the interaction of antibodies with oligosaccharide haptens (Zidovetzki et al. 1988) and hen egg white lysozyme (Zeder-Lutz et al. 1997).

Extrapolation of the curves shows that the Fab-antigen binding is driven by enthalpy below 37°C, by enthalpy and entropy above this temperature and (hypothetically) by entropy alone above 77°C.

Apparently, electrostatic interactions are the main driving force at lower temperatures. At higher temperatures, entropy increasing effects, such as the hydrophobic effect, can compensate entropy decreasing effects, e.g. stabilization of relative mobility, 'induced fit' or recruitment of additional water from solvent.

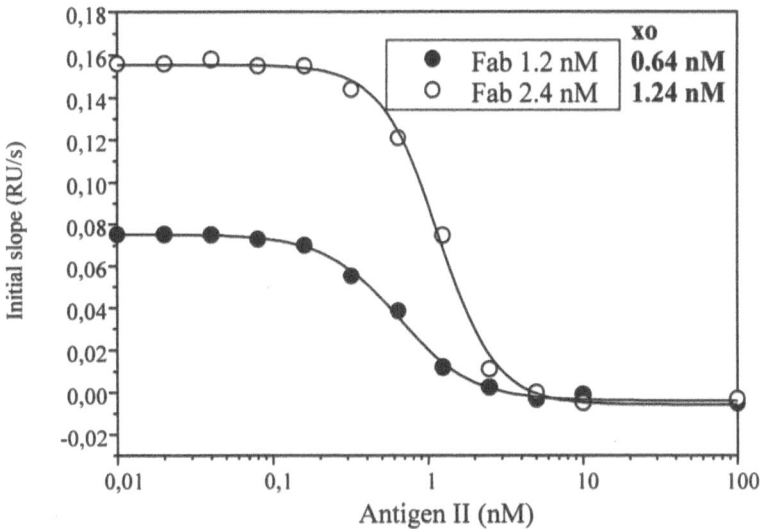

Figure 51: Standard curves for the Fab affinity in solution for antigen II

Figure 51 shows sensorgrams obtained with two different concentrations of Fab at a preincubation temperature of 5°C. Due to the low affinity (K_D = 0.04 nM) x_o is nearly proportional to the Fab concentration. The lowest feasible Fab concentration for this kind of experiment was about 0.4 nM, with an x_o of 0.24 nM antigen II. The time for an assay of preincubated samples was 15 min.

3.4.3.3 Ethanol effects on antigen binding

The affinity (and possibly) specificity of antibodies is influenced not only by temperature, but may also depend on other experimental conditions, such as buffer composition, pH and the presence of cosolvents (De Lauzon et al. 1994). Therefore, the binding of Fab to antigen I and antigen II in solution was measured in the presence of 0%, 10% and 20 % v/v ethanol in HBS buffer (Table 23). The affinity decreases for both antigens with increasing ethanol content.

Table 23: Ethanol effect on Fab binding to antigen I and II in solution at 20°C

Ethanol (vol.%)	Antigen	$x_o{}^a$ [nM]	$K_A{}^b$ [M^{-1}]	CR I[c] (%)	RA I[d] (%)
0	I	12.2	1.0E8	18.4	1.9
	II	2.24	5.3E9		
10	I	69.2	1.5E7	4.3	1.4
	II	3.0	1.1E9		
20	I	315	3.2E6	1.7	1.0
	II	5.3	3.1E8		

[a] x_o or IC_{50} is the antigen concentration at half maximum signal in sigmoidal curves, signal vs. log $C_{analyte}$
[b] $K_D = x_o - [Fab] / 2$; $K_A = 1/K_D$; Fab concentration was 3.6 nM
[c] CR I = crossreactivity for antigen I = x_o (II) / x_o (I).
[d] RA I = relative affinity = K_A (I) / K_A (II)

However, the crossreactivity was strongly changed in the presence of ethanol, in favor of antigen II. Therefore, it is possible to discriminate between both analytes in a mixture by comparing the immunoassay signals obtained in the presence or absence of ethanol, as was shown previously for triazine immunoassays (Stöcklein et al. 1997). It should be noted that the crossreactivity is defined as the ratio of analyte concentrations x_o causing 50% signal reduction in competitive (and related) assays. The K_D values calculated from x_o by the equation $K_D = x_o - [Fab]/2$ do not differ for the two analytes to the same extent as x_o. Therefore, the relative affinity of antigen I (RA I) was included in Table 23. It is evident that the observable change in crossreactivity depends on the Fab concentration. If x_o is close to [Fab/2], as for antigen II in this experiment, an relative increase in K_D corresponds to a smaller relative increase in x_o. If $x_o >> [Fab/2]$, both K_D and x_o will increase nearly to the same extent with increasing ethanol content.

3.4.3.4   pH effects on antigen binding

pH effects can be expected, as the crossreacting analytes differ with respect to ionizable groups (Weller et al. 1996). Therefore, Fab (5 nM) was preincubated with 100 nM antigen I or II at 20°C for 1h in HBS pH 7.4, or in analogous buffers, in which Hepes was replaced by citrate (pH 4 to 5.5) or glycine (pH 10). Free Fab was detected as in previous experiments and plotted against pH (Figure 52).

Whereas more than 99 % of the Fab was bound by free antigen II at all pH values used, antigen I binding was affected by pH values lower than pH 5.5. No binding was observed at pH 4.0. The breakpoint of the curve in Figure 52 matches the pk_a value of anilin (the antigen can be regarded as a substituted anilin). It can be concluded that the aromatic amino group of the antigen I must be deprotonated for binding to the Fab or Mab. In antigen II this group is substituted by the chloroacetyl residue and therefore not protonized at low pH. Hence, it is possible to eliminate the crossreactivity of analyte I just by lowering the pH from 7.4 to 4.0 in the preincubation step. However, as the Fab fragment stability is decreased at low pH, the preincubation time must be exact the same for each test.

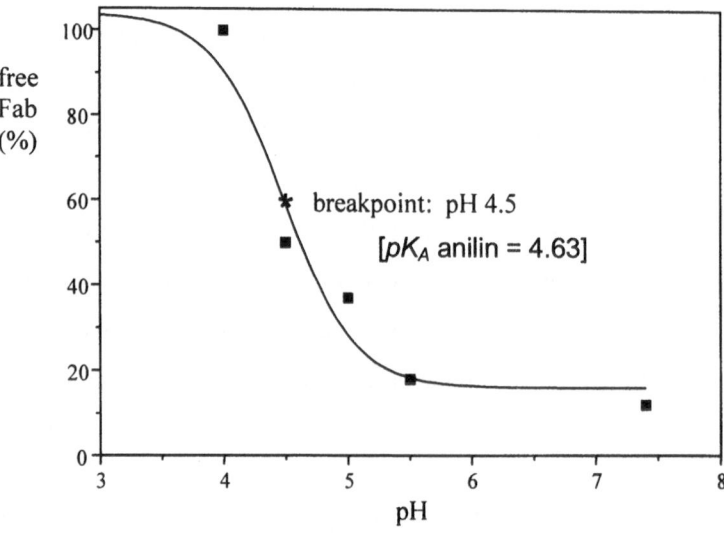

Figure 52: pH dependence of antigen I binding to Fab fragments

3.4.4   Conclusions

The binding properties of the monoclonal antibody and its Fab fragment were analyzed with BIACORE® 2000. Fab fragments were superior to the bivalent antibodies for the determination of rate constants. Antigen II (containing the coupling group) which was used for immunogen preparation and for immobilization, was bound by the Fab fragment with a 26 fold higher affinity than the actual target substance antigen I.

The binding reaction was driven by enthalpy below 37°C. Enthalpy-entropy compensation was observed. The affinity decreased and the lower limit of detection increased with increasing temperature or with increasing ethanol content of the binding buffer. The lower detection limit for antigen II ($K_D < 0.1$ nM) depended on the sensitivity of the SPR detection system, for antigen I ($K_D = 10$ nM) on the affinity.

The crossreactivity of analyte I decreased with increasing ethanol content or derceasing pH and was eliminated at pH 4.

Biospecific interaction analysis with BIACORE® 2000 can be used for rapid screening of pesticides in water samples. However, in the first place this instrument is a powerfool tool for the characterization and 'tuning' of antibodies prior to their use in immunoassays.

3.4.5   Acknowledgements

This work was supported by the Deutsche Forschungsgemeinschaft, Innovationskolleg INK 16 A1-1.

3.4.6   References

De Lauzon, S., Rajkowski, K.M., Cittanova, N. (1994): Investigation of a 17ß-estradiol-monoclonal antiestradiol antibody binding mechanism using dilute solutions of organic solvents. J. Steroid Biochem. Molec. Biol. 48, 225-233.

Friguet, B., Chaffotte, A.F., Djavadi-Ohaniance, L., Goldberg, M.E. (1985): Measurements of the true affinity constant in solution of antigen-antibody-complexes by enzyme-linked Immunosorbent assay. J. Immunol. Meth. 77, 305-319.

Karlsson, R., Fägerstam L., Nilshans, H., Persson, B. (1993): Analysis of active antibody concentration. Separation of affinity and concentration parameters. J. Immunol. Meth. 166, 75-84.

Karlsson, R. (1994): Real-time competitive kinetic analysis of interactions between low-molecular-weight ligands in solution and surface-immobilized receptors. Anal. Biochem. 221, 142-151.

Katmeh, M.F., Frost, G., Aherne, G.W., Stevenson, D. (1994): Development of an enzyme- linked immunosorbent assay for isoproturon in water. Analyst 119, 431-435.

Katmeh, M.F., Aherne, G.W., Stevenson, D. (1996): Development and evaluation of a chemiluminescent immunoassay for chlortoluron using a camera luminometer. Analyst 121, 329-332.

Krämer, P., Schmid, R.D. (1992): Biosensors for monitoring pesticides in water, pp. 1013-1021. In: Sensors - A Comprehensive Survey, Vol. 3 (Göpel, W., Hesse, J. and Zemel, J.N., eds.). VCH, Weinheim.

Lindner, W., Robey, F.A. (1987): Automated synthesis and use of N-chloroacetyl-modified peptides for the preparation of synthetic peptide polymers and peptide-protein immunogens. Int. J. Peptide Proten. Res. 30, 794-800.

Neri, D., Montigiani, S., Kirkham, P.M. (1996): Biophysical methods for the determination of antibody-antigen affinities. Trends Biotechnol. 14, 465-470.

Stöcklein, W.F.M., Warsinke, A., Scheller, F.W. (1997): Organic solvent modified enzyme linked immunoassay for the detection of triazine herbicides, pp. 373-381. In: ACS Symp. Ser. Vol. 657, Immunochemical technology for environmental applications (Aga, D.S. and Thurman, E.M., eds.). ACS, Washington.

Tijssen, P. (1985): Practice and theory of enzyme immunoassays, p. 124. Elsevier, Amsterdam.

Weller, M.G., Weil, L., Niessner, R. (1996): Analytica München, Poster P99

Wolter, R. (1995): Pflanzenschutzmittelfunde im Wasser, Forum Gewässerschutz und Pflanzenschutz, Industrieverband Agrar, Bonn, and: Umweltbundesamt Berlin-Information.

Zeder-Lutz, G., Zuber, E., Witz, J., Van Regenmortel, M.H.V. (1997): Thermodynamic analysis of antigen-antibody binding using biosensor measurements at different temperatures. Anal. Biochem. 246, 123-132.

Zidovetzki, R., Blatt, Y., Schepers, G., Pecht, I. (1988): Thermodynamics of oligosaccharides binding to a dextran-specific monoclonal IgM, Mol. Immunol. 25, 379-383.

3.4.7   Appendix - Definitions and formula

BIA: biospecific interaction analysis

CR: Crossreactivity of antibodies, defined as ratio of analyte concentrations causing 50% signal reduction in competitive assays. The best of the investigated analytes has 100% CR

ΔC_p (J mol^{-1} K^{-1}): heat capacity change. The plot DH vs. T gives a straight line with the slope Δc_p .

ΔG (kJ mol^{-1}): Gibbs free energy of binding, was calculated from the thermodynamic equation $\Delta G = -R * T * \ln K_A$; the gas constant R = 8.3145 J mol^{-1} K^{-1}; T is the abloute temperature in Kelvin

ΔH (kJ mol^{-1}): Enthalpy, was derived from the derivative ($-\Delta H/R$) of the regression curve of the van´t Hoff plot (Figure 50)

ΔS (J mol^{-1} K^{-1}): Entropy, was calculated from the equation $\Delta G = \Delta H - T*\Delta S$.

Fab: antigen-binding fragment (MW about 50 000) derived by papain cleavage of the IgG

k_{off} [s^{-1}]: measured dissociation rate constant. The data were fitted to the equation: $R' = R_0 e^{-k_{off}(t-t_0)}$; t is the time in seconds, t_0 the start time for dissociation, R_0 the response at start of dissociation (BIAevaluation software 2.1, BIACORE)

k_{on} [M^{-1} s^{-1}]: measured association rate constant. The data were fitted to the equation: $R' = R_{eq} e(1 - 1e^{-k_s(t-t_0)})$;

$k_s = k_{on} * C_{analyte} + k_{off}$; t is the time in seconds, t_o the start time for dissociation, R_{eq} the steady state response level. k_a was obtained from a secondary plot of k_s against C (BIAevaluation software 2.1, BIACORE)

K_A [M]: Association equilibrium constant = $1 / K_D$

K_D [M]: Dissociation equilibrium constant = $1 / K_A$

- from kinetic experiments: $K_D = k_{off} / k_{on}$

- from affiniy in solution: $K_D = x_o - b/2$, where x_o is the antigen concentration at half maximum signal, b is the concentration of binding sites.

The curve was corrected for bivalent antibodies by the following formula:

$s' = 1 - (1-s)^{1/2}$, where s is the relative initial response (between 0 and 1) at a given antigen concentration

Mab: monoclonal antibody (from mouse)

R′: Response in RU; R_{max}: maximum response

RU: resonance unit (dimensionless), 1000 RU correspond to 1 ng protein bound to 1 mm^2 or 6 mg protein / ml in solution

SPR: surface plasmon resonance

4 Monitoring of Toxic Effects

4.1 Endocrine Disruptors: Monitoring of Effects

Bertold Hock and Martin Seifert

Technical University of München at Freising-Weihenstephan, Department of Botany, D-85350 Freising, Germany

Abstract. Endocrine disruptors are exogenous natural or anthropogenic agents that produce adverse effects not only at the level of the individual, but also of the population and the community, by interfering with endogenous hormones in the body. Since these hormones are responsible for key processes in cellular control and communication as well as reproduction, development and behaviour, several severe impacts on ecosystems that occured during the last fifty years can now be explained by the presence of hormone mimics, acting as endocrine disruptors. Xenoestrogens such as DDT and PCBs are of special importance because they are leading to "feminized" wild life. Some of them are suspect of causing human sperm abnormalities, a decrease in sperm counts and a rise in hormone-related cancers. Since embryo development is particularly sensitive to estrogens, even small shifts in hormone levels experienced during the embryonic stage, e.g. by the presence of xenoestrogens, can cause malformations and cancer. The best known example is diethylstilbestrol (DES), a human "transplacental" carcinogen.

Since there are considerable structural differences between natural estrogens and xenoestrogens, the toxic potential cannot be derived from the chemical structure of individual substances, especially if xenoestrogens occur in complex mixtures and include yet unidentified compounds. Monitoring of effects takes advantage of the biological functions of estrogens, which act as signalling molecules. Components of the signal transduction chain can be used for effects monitoring, depending on the scope of the assay to be aimed at. The following options for detecting estrogen-like compounds are considered: receptor binding assays, DNA-binding tests of the receptor-ligand complex, and reporter gene assays measuring the estrogen-dependent modulation of gene activity. The perspectives for future biosensor approaches are discussed.

4.1.1   Introduction

Endocrine disruptors (EDs) have become a major issue during recent years. As the toxicological potential of a sample cannot be deduced from the chemical structure of individual substances, especially if as yet unidentified, active compounds are present, effects monitoring is required. There are several approaches to this task, ranging from bioassays to ED-responsive reporter gene and receptor assays.

This chapter includes a brief survey of the biological effects of estrogenic compounds as the major class of EDs and their signal transduction pathway. The main emphasis is laid upon the strategies available for the detection of xenoestrogens in environmental samples. It is expected that progress in biosensor technology will lead to major applications in the field of effects monitoring for the presence of EDs.

4.1.2   Hormone-disrupting chemicals in the environment

Several environmental disasters that occurred during the last fifty years are now related to EDs. This issue gained special attention as a result of the study of Colborn et al. (1993) on developmental effects of endocrine-disrupting chemicals in wildlife and humans. Public awareness was attained by the book "Our stolen future" of Colborn et al. (1996), who specifically warned of world-wide epidemics of feminized wild life, human reproductive system cancers, and declining male fertility. Table 24 lists the major incidents and their suspected causes.

Table 24: Survey on major incidents related to endocrine disruptors

Incident	Year	Interpretation	Selected references
Decline in the percentage of successful bald eagle nests (Florida, USA)	1947	sterility of 80% of the bald eagles	Broley (1958)
Decline of otters (England)	end of the 1950s	presence of organochlorines	Mason et al. (1986)
Decline of mink reproduction (Great Lakes, USA)	mid-1960s	PCBs in fish as suspected cause for reproductive failure	Dutton (1988)
Extra eggs in western gull nests (Southern California, USA)	early 1970s	female pairs share nests. DDT exposure suspected	Hunt and Hunt (1977), Fry and Toone (1981)
Decline of alligator hatching (Lake Apopka, Florida, USA)	1980	pesticide spill (1980): dicofol (Kelthane), an acaricide, containing DDE: gonadal deformities in male alligators	Woodward et al. (1993), Guillette (1994)
Increased incidence of breast cancer in women	main exposure during 1950-1972	xenoestrogens (organochlorines)	Davis et al. (1993)
Occupational oligospermia	1949	DDT	Singer (1949)
Human sperm abnormalities and declining sperm count	last 50 years	environmental factors suspected	Carlsen et al. (1992)

Fry et al. (1987) provided experimental evidence that environmental contaminants may cause reproductive problems. Eggs were taken from different gull colonies in relatively uncontaminated areas and injected with various EDs, among them DDT. Most amazingly, levels of DDT reported in contaminated areas were found to disrupt the sexual development of male birds. The presence of typically female cell types in the testicles or, in cases of higher doses, the formation of an oviduct clearly indicated a feminization of the males' reproductive tracts, although the animals had no visible defects and looked completely normal.

Severe consequences of endocrine disruption for the human reproductive system have been pointed out by Carlsen et al. (1992) who registered increasing human sperm abnormalities as well as a dramatic increase of testicular cancer during the last fifty years. A review of more than sixty studies, most from the USA and Europe, but also from Asia, South America and Africa resulted in the finding that average human male sperm counts had dropped by almost 50 percent between 1938 and 1990. Likewise, it was reported that organochlorines, especially DDT, may be a risk factor for breast cancer (Davis et al. 1993).

4.1.3   Embryo development is particularly sensitive to endocrine disruptors

The most intriguing consequences of endocrine disruption originate from prenatal effects. The influence of the hormonal environment of mouse embryos on the development and behaviour of adults was revealed by vom Saal (1989). It was found that differentiation of male fetuses into the masculine phenotype not only depends on the secretion of androgens from the testes, but is also influenced by differential exposure to estrogen during critical stages of prenatal development. Increased sexual activity of males in adult life were linked to higher levels of estrogen originating from neighboring female embryos (vom Saal et al. 1983). The males exposed to higher levels of estrogen had prostates that were fifty percent larger than those seen in brothers who had adjacent male fetuses in the womb.

Since critical stages of embryo development are particularly sensitive to estrogens, it is not surprising if synthetic estrogens and xenoestrogens can cause endocrine disruption. Diethylstilbestrol (DES) was the first synthetic agent specifically designed to have estrogenic activity. Although DES is not structurally similar to natural estrogens, it exhibits comparable biological functions. It was used for decades as a growth stimulant in cattle. As early as 1948, it was also prescribed to a large population of pregnant women to prevent miscarriages and other pregnancy complications. However, in 1971 the drug became associated with a rare form of vaginal cancer, clear-cell adenocarcinoma, detected in some of the adolescent daughters of women who had taken DES (Herbst and Bern 1981) and therefore became the first documented example of a human "transplacental" carcinogen, i.e. a chemical, which causes cancer in the daughter when given to the mother.

Likewise, exposure to estrogens *in utero* has been reported to affect male genital development in humans and mice. As adults, male mice exposed *in utero* to DES had a higher-than-average frequency of undescended testicles, testicular cancer, sperm abnormalities and prostate disease. Some of these outcomes were also reported for men exposed in utero to DES. The doses of DES required to cause malformations of the male reproductive tract were almost the same in mice and men. A detailed description of the cellular and molecular effects of developmental exposure to DES was given by Newbold (1995).

4.1.4   Monitoring strategies for estrogenic disruptors

These observations underline the necessity of monitoring programs for the presence of endocrine disruptors such as xenoestrogens. Figure 53 lists the chemical structures of natural and synthetic estrogens as well as important environmental contaminants that are known or suspected to modulate endocrine response pathways. Since there are considerable structural differences between natural estrogens and xenoestrogens, it is difficult to predict estrogenicity of xenobiotics on the basis of chemical structures. However, effects monitoring can measure endocrinological equivalents of even unknown substances.

The classical approach for assaying estrogenicity is the uterotropic response, which can be detected by an estrogen-induced increase in wet weight and tissue mass of the uterus. The advantages of this approach, e.g. by means of the mouse uterine bioassay, are obvious. It comes closest to the living situation and incorporates effects of metabolism, serum binding, and pharmacokinetics. However, large-scale screening requires faster and less expensive alternatives. These are found among *in vitro* assays.

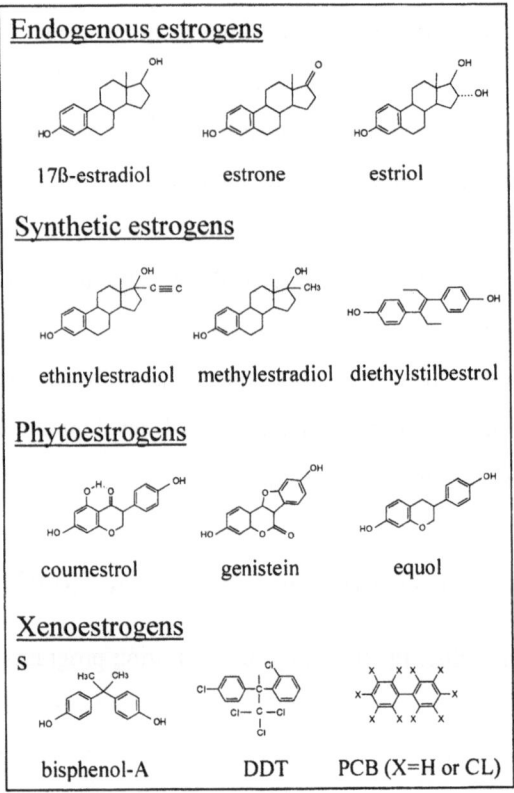

Figure 53: Chemical structures of estrogens and xenoestrogens

Estrogenic responses can be assayed *in vitro* at several levels of biological complexity (cf. Korach and McLachlan 1995): (1) Proliferation of estrogen-responsive cells, such as MCF-7 breast cancer cells. The E-SCREEN by Soto et al. (1992) assays the cell yield achieved after 4-6 d of culture supplemented with 5-10% charcoal-dextran stripped human serum in the presence or absence of estrogens.

(2) Measurement of biomarkers such as vitellogenin (a generic, ancient lipoprotein, synthesized in the liver under the control of estrogen and taken up into the oocyte, cf. Sumpter and Jobling 1995), and lactoferrin (an iron-binding glycoprotein belonging to the transferrin gene family with antibacterial and antiviral activity; high levels are found in neutrophils, lactating mammary glands and the uterus, cf. Teng 1995. Estrogen-controlled biomarkers are covered in more detail by Hansen et al. (1997) in this volume.

(3) Use of components of the estrogen signal transduction chain. This signalling pathway comprises (a) estrogen binding by the estrogen receptor (ER), a member of a family of transcription factors, (b) the interaction of the hormone-bound ER with estrogen-responsive DNA elements (ERE), and (c) alteration of the transcription efficiency of ERE-containing genes, among them the vitellogenin, lactoferrin and epidermal growth factor genes.

4.1.5 Receptor binding assays

Receptors can be used as binding proteins for pharmacologically or toxicologically relevant substances in unknown samples. Receptor binding assays basically use the same approach as immunoassays, which employ antibodies as binding proteins. Whereas antibody binding is not linked to any biological effect of the ligand, receptor binding implies a function as agonist or antagonist. We have recently developed a non-radioactive receptor binding assay (Figure 54a). Measurements are carried out in 96-well microwell plates. First, a estradiol-BSA conjugate is adsorbed to the walls of the microwells. In the second (competition) step an estradiol solution of defined concentration is incubated together with the recombinant human estradiol receptor

(A)

(a) Assay with immobilised receptor

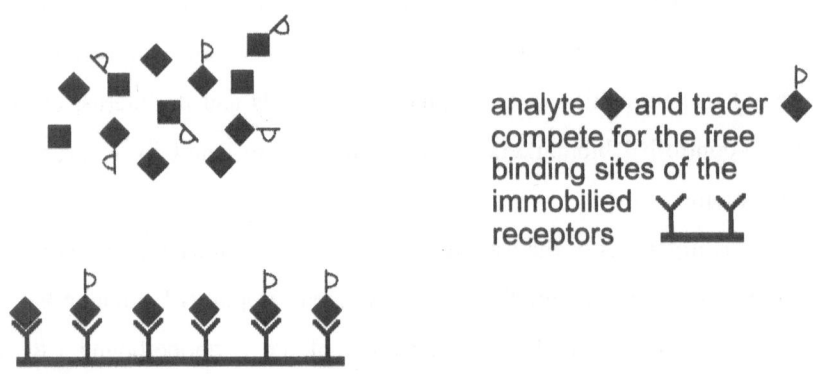

analyte ◆ and tracer ◆
compete for the free
binding sites of the
immobilied
receptors

(b) Assay with immobilised coating conjugate

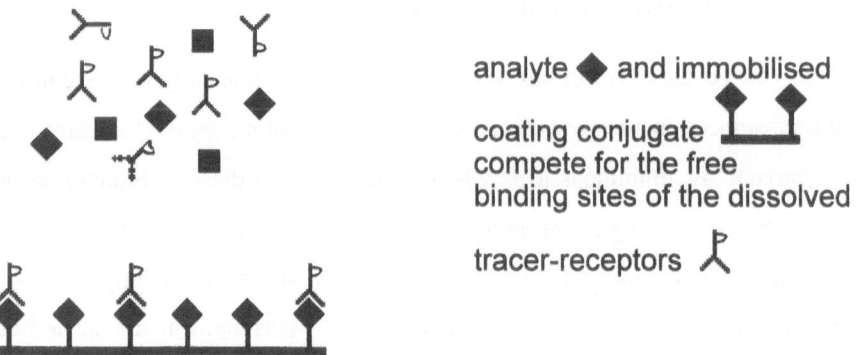

analyte ◆ and immobilised

coating conjugate
compete for the free
binding sites of the dissolved

tracer-receptors

Figure 54:   Receptor assays for estrogens. (a) General scheme for an enzyme-linked receptor assay.
(b) Estradiol calibration curve

(B)

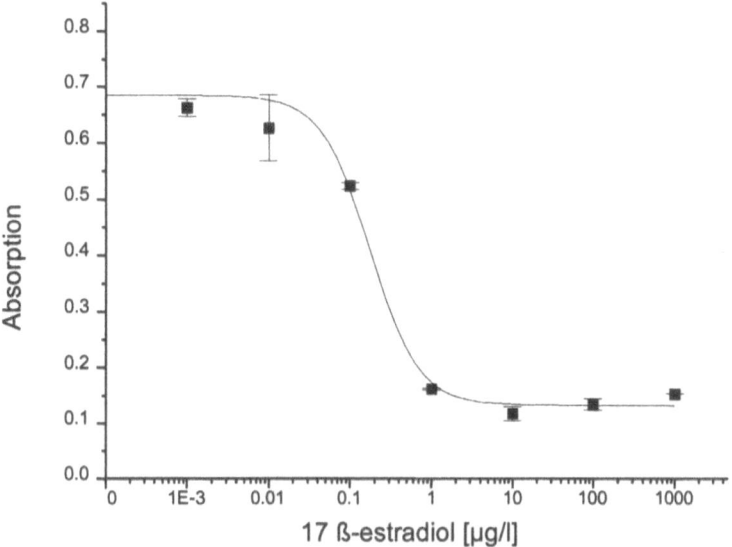

Figure 54, continued

(form α). After the receptor binding reaction, a biotinylated mouse-anti-human estrogen receptor antibody is added. With a streptavidin-biotin enhancement system an inverse relation between enzyme activity and effect-concentration (estradiol) is observed (Figure 54b). This enzyme-linked receptor assay (ELRA) presently shows a detection limit of 0.1 μg/l for 17-ß-estradiol (Figure 54b). Oosterkamp et al. (1996) reached a detection limit of 1.3 μg/l for 17-ß-estradiol using reversed phase liquid chromatography coupled to receptor affinity detection. A commercially available homogen estrogen screening assay (PanVera 1997) is based on a fluorescence polarisation instrumentaion shows a detection limit for 17-ß-estradiol of 1.7 μg/l (derived from the calibartion curve; 80% competitor binding).

4.1.6   DNA binding assays with steroid hormone receptors

The estrogen receptors as hormone-inducible transcription factors regulate the expression of several genes, which take part in growth and differentiation of tissues. They expose two conserved domains (Figure 55a): (1) a DNA binding domain with approximately 70 aminoacid residues, which folds around two zink atoms. The estrogen receptors as most of the members of the steroid superfamily recognize a common DNA consensus sequence with 6 base pairs and bind as dimers to the regulatory regions of specific genes. The relative distance and arrangement of the two hexamers determines the specificity. (2) The hormone recognition domain consists of approximately 200 aminoacid residues and binds estrogens as well as xenoestrogens. This leads to a conformational change of the ligand binding domain, followed by dimerization of the receptor, binding to the regulatory regions of specific genes, and activation or repression of gene expression. It is believed that the longer the receptor-ligand complex remains attached to the estrogen-response element, the longer the complex modulates gene activity. Once the complex is removed, gene regulation also ceases. Gene expression is regulated only in the dimeric form.

Since DNA binding only occurs to a significant extent when the hormone binding sites are occupied by suitable ligands, this event can be used for effects monitoring. Cheskis and Freedman (1996) have applied this principle for a protein-DNA binding assay with the retinoid X receptor (RXR), which also belongs to the steroid receptor family. The natural ligand is 9-cis-retinoic acid. RXR plays an important role in hormonal signal transduction because it forms heterodimers with several nuclear receptos including the vitamin D_3 receptor (VDR). Heterodimerization increases binding and transactivation of DNA response elements by receptors. Several ligands modulate the dimerization state.

Cheskis and Freedman (1996) applied the BIAcore biosensor system for their studies. First, streptavidin was immobilized to the sensor surface, followed by

(a) Receptor protein

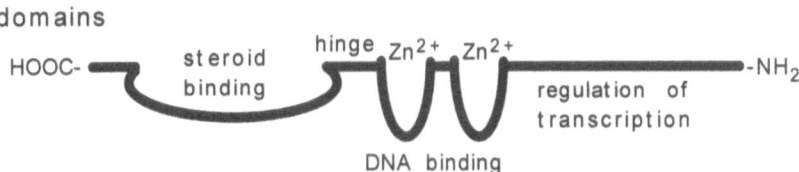

domains

HOOC- ⎰ steroid binding ⎱ hinge Zn²⁺ Zn²⁺ -NH₂

regulation of transcription

DNA binding

(b) DNA-binding assay

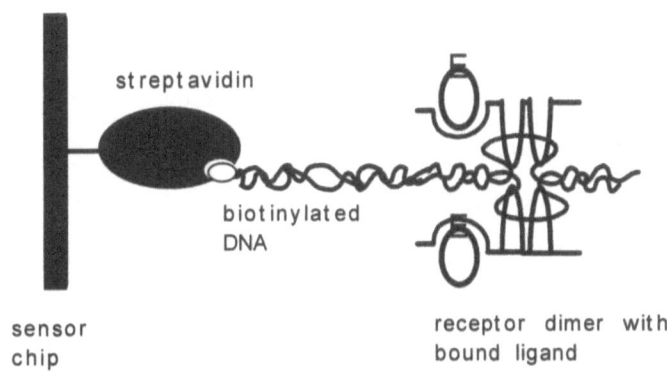

streptavidin

biotinylated DNA

sensor chip

receptor dimer with bound ligand

Figure 55: Estrogen receptor and ist use for DNA binding assays. (a) Domain structure of the estrogen receptor. (b) Protein-DNA binding assay for the detection of natural ligands and their analoga

biotinylated DNA (Figure 55b). For this purpose, an oligonucleotide duplex with a specific VDR response element was used. It contains a relatively strong binding site for VDR-RXR heterodimers in vitro. It could be shown that components of the signal transduction chain, lipophilic ligands such as 1,2,5-dihydroxy-vitamin D_3 and 9-cis-retinoic acid, can modulate protein-DNA interactions. Whereas 1,2,5-dihydroxy-vitamin D_3 increases the VDR-RXR binding to the VDR response element, 9-cis-retinoic acid has the contrary effect. This approach enables the examination of synthetic hormone analogues as well as xenohormones. It can be expected that this principle will be applied for further nuclear hormone receptors.

4.1.7 Reporter gene assays

Progress in recombinant technologies has speeded up the construction of estrogen-inducible expression systems. cDNA clones for the human estrogen receptor (hER) first became available by Walter et al. (1985), followed by expression systems for the estrogen receptor,e.g. *E. coli* (Wittcliff et al. 1990) or yeast (McDonnell et al. 1991). Finally, reporter gene assays could be developed on this basis. The best known estrogen-responsive expression system is the yeast estrogen screen (YES), Figure 56 by Arnold et al. (1996a). It was generated by introducing two plasmids into a yeast strain, an expression plasmid including the gene for the human estrogen receptor and a reporter plasmid containing two estrogen response elements linked to the lacZ gene. The yeast cells respond to estrogen exposure by the expression of ß-galactosidase. The enzyme activity can be measured by the conversion of nitrophenyl ß-D-galactopyranosid to nitrophenol.

When 17-ß-estradiol, diethylstilbestrol, or the xenoestrogens o,p'-DDT and octyphenol were added, a significant increase of ß-galactosidase activity was observed, but not with the anti-estrogens ICI164,384, dexamethason or the hormone testosteron. If the recombinant yeast strain was grown in the presence of human albumin or steroid hormone-binding globuline, the addition of estradiol led to a considerably diminished ß-galactosidase activity. This demonstrates the reduced bioavailability of estrogen in presence of these serum components. Most important, this does not apply to xenoestrogens, which are considerably less bound by serum.

Arnold et al. (1996b) claimed that the YES system could be applied for the detection of synergistic effects of natural and environmental estrogens. The weakly active pesticides dieldrin, toxaphen or endosulfan led to synergistic effects, if two of these compounds were combined. These findings triggered a highly controversial discussion since the results could not be confirmed (Ramamoorthy et al. 1997a, Ramamoorthy et al. 1997b, Ashby et al. 1997).

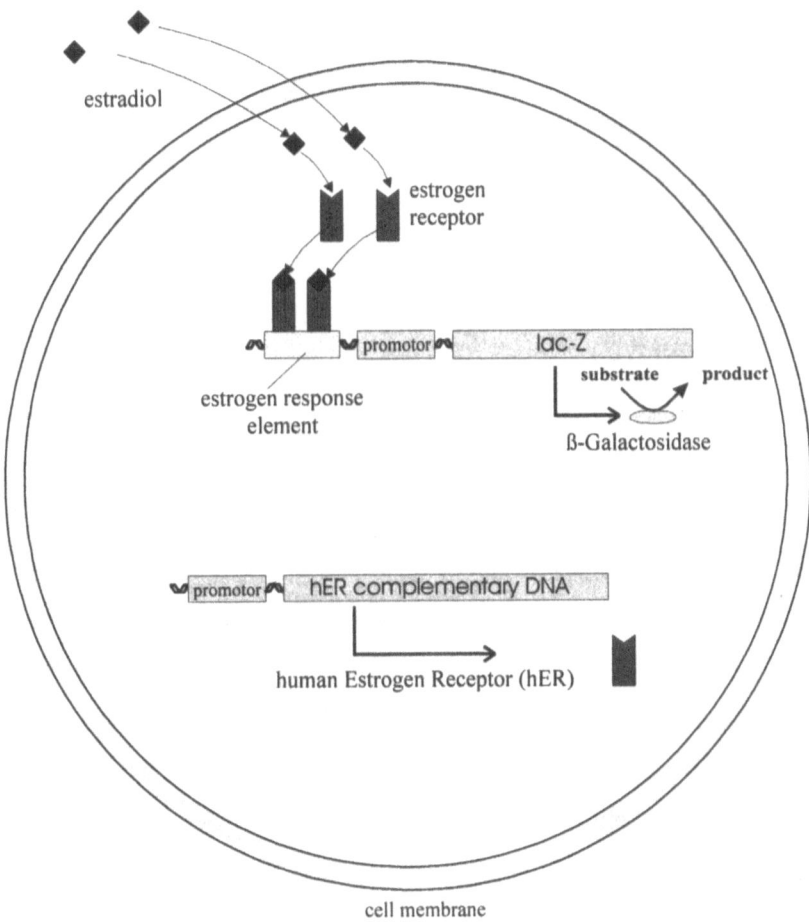

Figure 56: The yeast estrogen screen

ß-Galactosidase as a reporter system for gene expression is only one of several options. A promising alternative for biosensors is the use of luciferases. Selifonova et al. (1993) applied *E. coli* for the construction of Hg (II) detectors. Exposure to the heavy metal triggered bioluminescence. In this case the mercury resistence operon (mer) was ligated with the promoterless luxCDABE gene from *Vibrio fisheri*, cloned into the plasmid pRB28 and used for the transformation of *E. coli*. In this case a Hg (II) sensitivity of 1nM was reached, in the presence of the Hg (II) uptake system even 0.5 nM. This bioluminescence sensor was not only sensitive but also selective.

The application of this principle for the detection of xenoestrogens appears to be very attractive. Indeed there already exists a luciferase assay using transformed MCF-7 breast cancer cells (Klotz et al. 1996) for the detection of estrogenic chemicals.

4.1.8   Perspectives

The relevance of endocrine disruptors and growing public concern is mirrored by the approval of the U.S. Congress in 1996 in line with the legislation that U.S. laboratories will soon carry out large-scale testing of pesticides and other potentially hazardous chemicals, individually and in mixtures, to determine their effects on endocrine systems. The U.S. Environmental Protection Agency (EPA) created an Endocrine Disruptors Screening and Testing Advisory Committee to establish a framework for identifying chemicals that pose hazards to human or wildlife endocrine systems. A research inventory (Endocrine Disruptor Research Inventory: EDRI) was constructed in parallel with the framework. According to legislation, the EPA must develop a screening and testing program within two years, implement the program within three years and report the congress four years. A research inventory (Endocrine Disruptors Research Initiative: EDRI) was constructed in parallel with the framework.

The concepts for effects monitoring outlined in this paper are characterized by different levels of complexity and power. Receptor binding assays apply to both, agonists and antagonists, but do not discriminate between these different types of ligands. DNA binding assays and reporter gene assays basically respond only to agonists by an increase of DNA binding or expression of the reporter gene, respectively. However, antagonists can be recognized in the presence of defined levels of estrogen by an attenuation of the signal. Regarding mixtures of agonists and antagonists the effects may be cancelled in DNA binding and reporter gene assays, depending on the respective concentrations. Signals obtained by receptor binding assays only depend on the affinities of the potential ligands.

It is not clear, yet, to which extent synergistic effects can be recorded by DNA binding and reporter gene assays. If the estrogen receptors are equipped with allosteric binding or phosphorylation sites (Arnold et al. 1997), an amplification of estrogenic effects by the presence of two different ligands could be detected by these assays. So far, clear evidence for this possibility is not available.

The three approaches to effects monitoring based on different components of the signal transduction chain cannot recognize estrogenic effects due to other interferences, such as effects on synthesis, metabolism and transport of endogenous estrogens as well as superior hormone centres. As a consequence, suborganismic tests described above can only be considered as screening methods for the presence of endocrine disruptors and should be accompanied by a justifiable extent of biological assays.

A distinct disadvantage of effects monitoring is the lack of chemical identification and structural analysis. This lack can be relieved by coupling the both principles. Two alternatives are feasible; they differ by the sequence of their components. The first possibility is the separation of analytes in environmental samples by conventional chromatographical methods, followed by effects monitoring. This approach has been used by Oosterkamp et al. (1996). They applied reversed phase liquid chromatography, coupled on-line to receptor affinity detection with fluorescent estrogens as tracers. (B) The alternative followed in our group proceeds in the reversed order. Receptor chromatography for the preconcentration of endocrinic substances from environmental samples followed by conventional analytics (GS-MS, LC-MS, NMR).

Considering the extensive tasks and challenges to be met by future effects monitoring, it is obvious that a high degree of automation will be required, especially if monitoring of several parameters has to be carried out simultaneously. This demand is already clear in the field of estrogenic disruptors, where different levels of test should be integrated within a self-contained measuring device. It becomes even more important by when one considers that endocrine disruption concerns not only processes linked to the estrogen receptors but also to other receptors, such as androgen, thyroid and gestagen receptors.

The solution to these analytical problems are biosensors, i.e. self-contained integrated systems which are capable of providing quantitative or semiquantitative analytical information using a biological recognition element which is directly spatially coupled to a transduction element (cf. Cammann 1997, this volume). Although chemical sensing basically differs from effect monitoring in its scope, the construction elements are fundamentally the same for both chemical sensors and biosensors. Whereas chemical sensors in the most simple case generate signals which are converted into analyte concentrations, biosensors used for effects monitoring provide pharmacological and/or toxicological equivalents. It is believed that the field of effects monitoring will benefit in the future from the already established highly advanced biosensor technologies, especially if receptor binding and DNA binding assays can be integrated into suitable sensor arrays.

4.1.9   Acknowledgements

We would like to thank the BMBF for financial support (02-BU9647/0). We are grateful to Stefanie Rauchalles for typing the manuscript.

4.1.10   References

Arnold, S.F., Klotz, D.M., Collins, B.M., Vonier, P.M., Guillette, Jr., L.J., McLachlan, J.A. (1996b): Synergistic activation of estrogen receptor with combinations of environmental chemicals Science 272, 1489-1492.

Arnold, S.F., McLachlan, J.A. (1996c): Environmental estrogens. American Scientist Internet Publication.

Arnold, S.F., Melamed, M., Vorjojeikina, D.P., Notides, A.C., Sasson, S. (1997): Estradiol-binding mechanism and binding capacity of the human estrogen receptor is regulated by tyrosin phosphorylation. Mol. Endocrinol. 11, 48-53.

Arnold, S.F., Robinson, M.K., Notides, A.C., Guillette, Jr., L.J., McLachlan, J.A. (1996a): A yeast estrogen screen for examining the relative exposure of cells to natural and xenoestrogens. Environ. Health Perspect. 104, 544-548.

Ashby, J., Lefevre, P.A., Odum, J., Harris, C.A., Routledge, E.J., Sumpter J.P. (1997): Synergy between synthetic estrogens. Nature 385, 494.

Broley, C. (1958): The plight of the American bald eagle. Audubon Magazine 60, 162-163.

Camman K. (1997): Stability of biosensors for environmental monitoring. In: *this volume*.

Carlsen, E., Giwercman, A., Keiding, B., Skakkebaek, N. (1992): Evidence for decreasing quality of semen during past 50 years. British Med. J. 305, 609-613.

Cheskis, B., Freedman, L.P. (1996): Modulation of nuclear receptor interactions by ligands: kinetic analysis using surface plasmon resonance. Biochemistry 35, 3309-3318.

Colborn, T., Dumanoski, D., Myers, J.P. (1996): Our stolen future. Dutton, New York.

Colborn, T., Vom Saal, F.S., Soto, A.M. (1993): Developmental effects of endocrine-disrupting chemicals in wildlife and humans. Environ. Health Perspect. 101, 378-384.

Davis, D.L., Bradlow, H.L., Wolff, M., Woodruff, T., Hoel, D.G., Anton-Culver, H. (1993): Medical hypothesis: xenoestrogens as preventable causes of breast cancer. Environ. Health Perspect. 101, 372-377.

Dutton, D. (1988): Worse than disease: Pitfalls of medical progress. Cambridge University Press.

Fry, D., Toone, C., Speich, S., Peard, R. (1987): Sex ration skew and breeding patterns of gulls: Demographic and toxicological considerations. Studies in Avian Biology 10, 26-43.

Fry, D., Toone, M. (1981): DDT-induced feminization of gull embryos. Science 213, 922-924.

Guillette, L.J. (1994): Developmental abnormalities of the gonad and abnormal sex hormone concentrations in juvenile alligators from contaminated and control lades in Florida. Environ. Health Perspect. 102, 680-688.

Hansen, P.-D., v. Usedom, A., Dizer, H., Hock, B., Marx, A., Sherry, J., McMaster, M. (1997): Vitellogenin as biomarker for endocrine disruptors. In: *this volume*.

Herbst, A.L., Bern, H.A. (1981): Developmental effects of diethylstilbestrol (DES) in pregnancy. Thieme-Stratton, New York.

Hunt, G., Hunt, M. (1977): Female-female pairing in western gulls (Larus occidentalis) in southern California. Science 196, 1466-1467.

Klotz, D.M., Beckmann, B.S., Hill, S.M., McLachlan, J.A., Walters, M.R., Arnold, S.F., (1996): Identification of environmental chemicals with estrogenic activity using a combination of in vitro assays. Environ. Health Perspect. 104, 1084-1089.

Korach, K.S., McLachlan, J.A. (1995): Techniques for detection of estrogenicity. Environ. Health Perspect. 103 (Suppl. 7), 5-8.

Mason, C., Ford, T., Last, N. (1986): Organochlorine residues in British otters. Bull. Environm. Contamination and Toxicol. 36, 656-651.

McDonnell, D.P., Nawaz, Z., Densmore, C., Weigel, N.L., Pham, T.A., Clark, J.H., O'Malley, B.W. (1991): High level expression of biologically active estrogen receptor in saccharomyces cervesiae. J. Steroid Biochem. Molec. Biol. 39, 291-297.

Newbold, R. (1995): Cellular and molecular effects of developmental exposure to diethylstilbestrol: Implications for other environmental estrogens. Environ. Health Perspect. 103 (Suppl. 7): 83-87.

Oosterkamp, A.J., Villaverdeherraiz, M.T., Irth, H., Tjaden, V.R., Van der Greef, J. (1996): Reversed-phase liquid chromatography coupled on-line to receptor affinity detection based on the human estrogen receptor. Anal. Chem. 68, 1201-1206.

Product information PanVera Corporation(product number P2313): Estrogen Competitor Screening Kit (1997).

Ramamoorthy, K., Wang, F., Chen, I.C., Norris, J.D., McDonnell, D.P., Leonard, L.S., Gaido, K.W., Bocchinfuso, W.P., Korach, K.S., Safe, S. (1997): Estrogenic activity of a dieldrin/toxaphene mixture in the mouse uterus, MCF-7 human breast cancer cells, and yeast-based estrogen receptor assays: no apparent synergism. Endocrinology 138, 1520-1527.

Ramamoorthy, K., Wang, F., Chen, I.-C., Safe, S., Norris, J.D., McDonnel, D.P., Gaido, K., Bocchinfuso, W.P., Korach, K.S. (1997): Potency of combined estrogenic pesticides, Science 275, 405-406.

Selifonova, O., Burlage, R., Barkay, T. (1993): Bioluminescent sensors for detection of bioavailable Hg(II) in the environment. Appl. Environ. Microbiol. 59, 3083-3090.

Singer, P.L. (1949): Occupational oligospermia. JAMA 140, 1249.

Soto, A.M., Lin, R.-M., Justicia, H., Silvia, R.M., Sonnenschein, C. (1992): An "in culture" bioassay to assess the estrogenicity of xenobiotics, p. 295-309. In:

Chemically-induced alterations in sexual development: The wildlife/human connection (Colborn, T., Clement, C. eds.) Princeton Sci. Publ., Princeton, N.J.

Sumpter, J.P., Jobling, S. (1995): Vitellogenesis as a biomarker for estrogenic contamination of the aquatic environment. Environm. Health Perspect. 103 (Suppl. 7), 173-178.

Teng, C. (1995): Mouse lactoferrin gene: A marker for estrogen and epidermal growth factor. Environm. Health Perspect. 103 (Suppl. 7), 17-20.

vom Saal, F., Grant, W.M., McMullen, C.W., Laves, K.S. (1983): High fetal estrogen concentrations: correlation with increased adult sexual activity and decreased aggression in mice. Science 220, 1306-1309.

vom Saal, F. (1989): Sexual differentiation in litter baring mammals: influence of sex of adjacent fetuses in utero. J. Animal Sci. 67, 1824-1840.

vom Saal, F., Bronson, F. (1980): Sexual characteristics of adult female mice are correlated with their blood testosterone levels during prenatal development. Science 208, 597-599.

Walter, P., Green, S., Green, G., Krust, A., Bornert, J.M., Jeltsch, J.M., Staub, A., Jensen, E., Scarce, G., Waterfield, M., Chambon, P. (1985): Cloning of the human oestrogen receptor cDNA. Proc. Natl. Acad. Sci. 82, 7889-7893.

Wittcliff, J.L., Wenz, L.L., Dong, J., Nawaz, Z., Butt, T.R. (1990): Expression and characterisation of an active human estrogen receptor as a ubiquitin fusion protein from Escherichia coli. The Journal of Biological Chemistry 265, 22016-22022.

Woodward, A., Percival, H., Jennings, M., Moore, C. (1993): Low clutch viability of American alligators on Lake Apopka. Florida Science 56, 52-63.

5 Monitoring of Genotoxicity

5.1 Recombinant *Escherichia Coli* Cells as Biodetector System for Genotoxins

Gerda Horneck[1], Leonid R. Ptitsyn[2], Petra Rettberg[1], Olga Komova[3], Stanislav Kozubek[4], Eugene A. Krasavin[3]

[1]Deutsche Forschungsanstalt für Luft- und Raumfahrt, Institut für Luft- und Raumfahrtmedizin, Abteilung Strahlenbiologie, D-51170 Köln, Germany

[2]State Scientific Center of Russian Federation, GNII Genetica, 1,1-st Dorozhny Proezd, 113545 Moscow, Russia

[3]Joint Institute for Nuclear Research, Division of Radiation and Radiobiological Research, 141980 Dubna, Moscow Region, Russia

[4] Institute of Biophysics, ASCR, Brno, Czech Republic

Abstract. A bacterial biodetection system has been developed for rapid detection of environmental genotoxins. This cellular bioassay is based on the receptor reporter principle with the SOS system as receptor sensitive to DNA damage and the bioluminescence system as rapid optical reporter. Since the bioassay combines the SOS system with the bioluminescence system, we have termed it "SOS *lux* test". For the SOS *lux* test, a recombinant plasmid pPLS1 (DSM 10333) was constructed in which the promoterless operon of bioluminescence of *Photobacterium leiognathi* 54D10 (*lux C,D,A,B,F,E*) (Krasnojarsk Institute of Biophysics Collection) is under control of a SOS promoter (part of the *cda* gene of the plasmid ColD, which carries a strong SOS promoter). This plasmid pPLS1 can be used to transform any *Escherichia coli recA$^+$* system or other microorganisms with a SOS system, suitable for the detection of a specific genotoxic agent or combinations of genotoxic agents. The test system consists of recombinant bacterial cells in log-phase with pPLS1 that in response to the activity of a DNA damaging agent (e.g. chemicals, biotoxins or radiation) induce their SOS system and thereby the expression of the bioluminescence operon. The level of SOS response is quantified by measuring the light emission of the cells during their growth after treatment with the genotoxin. Compared with other bacterial bioassays based on the SOS response, such as the SOS chromotest or the *umu* test, the SOS *lux* test shows similar sensitivity.

5.1.1   Introduction

It has been well established that increasing levels of environmental pollution pose a health hazard to humans and ecosystems. To identify such sources of potential hazard, biological systems have been developed complementary to chemical and physical detection methods (Scheller and Schmid 1991). For genotoxicity assessment, bacterial test systems have been established that response to damage to their DNA induced by the genotoxin. These test systems are mainly based on the ability of certain bacteria to invoke the SOS regulon in response to DNA damaging agents. The most commonly used bacterial genotoxicity tests based on the SOS system are the SOS chromotest (Quillardet et al. 1982) and the *umu* test (Oda et al. 1985).

Within the project "On-line monitoring of mutagenic and carcinogenic agents by SOS dependent LUX test" of the program "Cooperation in Science and Technology with Central and Eastern European Countries" of the European Commission (Contract no. CIPA CT-94-0122), the SOS *lux* test was further developed and the response kinetics were determined for different chemicals and UV and γ radiation. It was demonstrated that the *lux* gene expression is detected within one to two hours after treatment with the genotoxin under investigation and that the sensitivity of the SOS *lux* test is comparable to that of other known genotoxicity bioassays.

5.1.2 The SOS *lux* bioassay

5.1.2.1 The SOS receptor system

Genotoxic agents (e.g. chemicals or radiation) induce a variety of genotoxin specific DNA injuries. Depending on their nature and position in the DNA, the lesions may

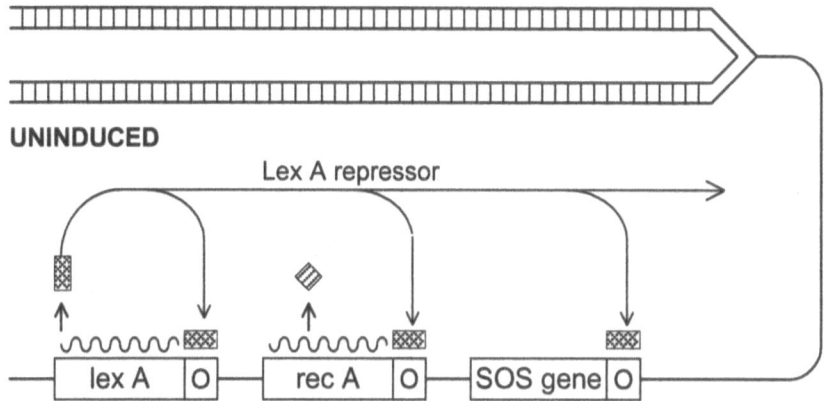

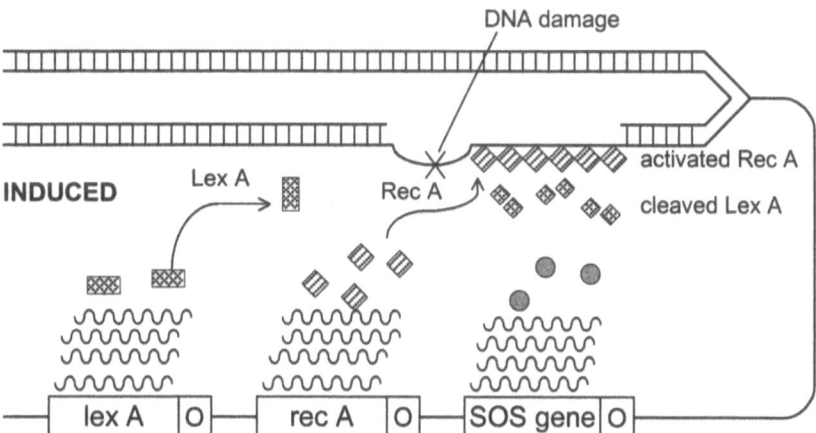

Figure 57: Simplified scheme of the SOS system in *E. coli* (modified from Friedberg 1985)

have different implications as follows: (i) the lesion may be excised or reverted with no further consequences for the cell, (ii) the lesion may induce the SOS response with error-prone repair pathways leading to mutagenesis, or (iii) the lesion remains unrepaired and leads to cell death. For the regulation of the SOS system, the two SOS genes *lexA* and *recA* are essential. The LexA protein acts as a repressor for all SOS genes, including *lexA* and *recA*. Damage to the DNA releases the SOS signal, probably through single stranded DNA (Sassanfar and Roberts 1990) which activates the RecA protein. The activated RecA protein inactivates the LexA repressor by promoting its autodigestion reaction. As a consequence, a cascade of physiological phenomena, the so-called SOS response, is induced (Figure 57). This includes the synthesis of a number of proteins involved in mutagenesis, such as RecA and UmuC/D (Witkin 1976).

SOS dependent bacterial test systems make use of this fact that in response to DNA damaging agents the diverse set of SOS functions is induced. In the SOS chromotest, *E. coli* PQ37 cells with the structural genes for β-galactosidase, *lacZ*, under the control of a SOS controlled gene, *sulA*, are used as test system for genotoxicity (Quillardet et al. 1982, 1989, Quillardet and Hofnung 1993, White and Rasmussen 1996). The *umu* test makes use of a recombinant *Salmonella typhi-murium* TA 1535 (pSK1002) strain with a fused *umuC:lacZ* operon carried on the plasmid (Oda et al. 1985, 1995, Nakamura et al. 1987, Reifferscheid et al. 1991, ISO CD 13829 1996). In both systems, the SOS induction potency is determined from a colorimetric assay for β-galactosidase synthesized in response to a genotoxin. The lower limits of detection of genotoxins are in the nanomolar to micromolar range (McDaniels et al. 1990, Quillardet and Hofnung 1993). Comparing the SOS genotoxic potency with the mutagenic potency, as determined by the Ames test (Ames et al. 1973, Maron and Ames 1983), for 82 % of the chemical compounds tested similar responses are observed in both tests (Mersch-Sundermann et al. 1994, Quillardet and Hofnung 1993). It was further shown for 65 confirmed carcinogens - according to the classification by Lewis (1991), that the capacity to detect carcinogens was 62 % with the SOS chromotest and 65 % for 44 suspected carcinogens (Quillardet and Hofnung 1993).

5.1.2.2   The LUX reporter system

Bioluminescent bacteria are found predominantly in the marine environment. Their bioluminescence system, the *lux* system, involves the genes *luxA* and *luxB* which code for the two nonidentical luciferase subunits α and β, and the genes *luxC*, *luxD* and *luxE* which code for the three proteins of a fatty acid reductase complex. In certain photobacteria, an additional gene *luxF* is found (Hastings et al. 1985). The bioluminescence reaction mechanism involves the oxidation of a long-chain fatty aldehyde and reduced flavin mononucleotide $FMNH_2$. The reaction products are a long-chain fatty acid and oxidized flavin. The reaction

$$FMNH_2 + RCHO + O_2 \rightarrow FMN + H_2O + RCOOH$$

leads to the emission of blue-green light with a peak at 490 nm. The bioluminescence reaction is outlined in Figure 58. The *lux* genes are arranged in the *lux* operon as follows: *luxCDAB(F)E*.

The *lux* gene system has been used in several instances as a reporter for investigating gene expression in bacteria. The main advantage is the availability of a real-time, *in-vivo* and non-disruptive monitoring system. Genetically controlled bacterial luminescent biosystems have been developed to detect a variety of environmental pollutants, as reviewed by Stewart et al. (1991) and Chatterjee and Meighen (1995). They are characterized by the coupling of a receptor component (e.g. promoters that are subject to environmental regulation) with the genes required for bioluminescence as reporter component. Several recombinant bacteria with plasmids carrying the promoterless *lux* operon (*luxCDABE*) or only the *luxAB* genes under control of a special inducible promoter have been developed. Examples are the detection of mercury (Selfinova et al. 1993, Tescione and Belfort 1993), arsenic and cadmium (Corbisier et al. 1993) and naphthalene and salicylate (Burlage et al. 1990,

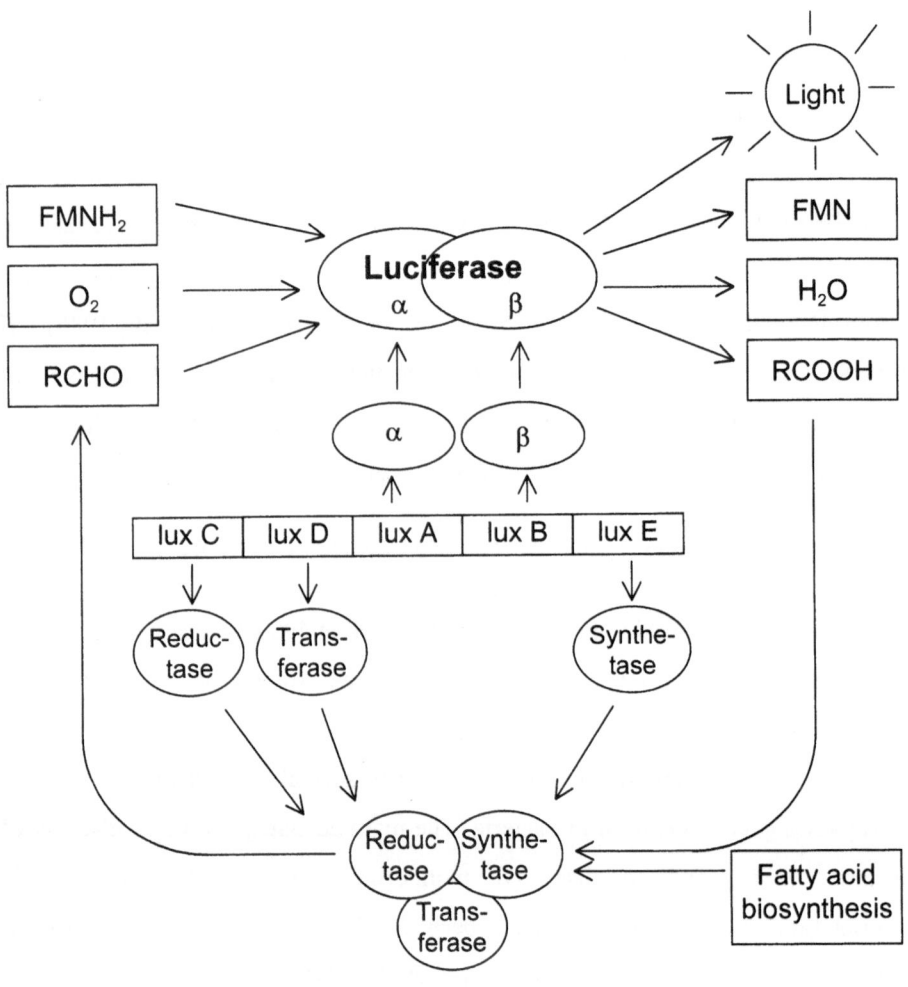

Figure 58: Scheme of the bacterial bioluminescence reaction and the relationship of the *lux* genes with the corresponding proteins (modified after Chatterjee and Meighen 1995)

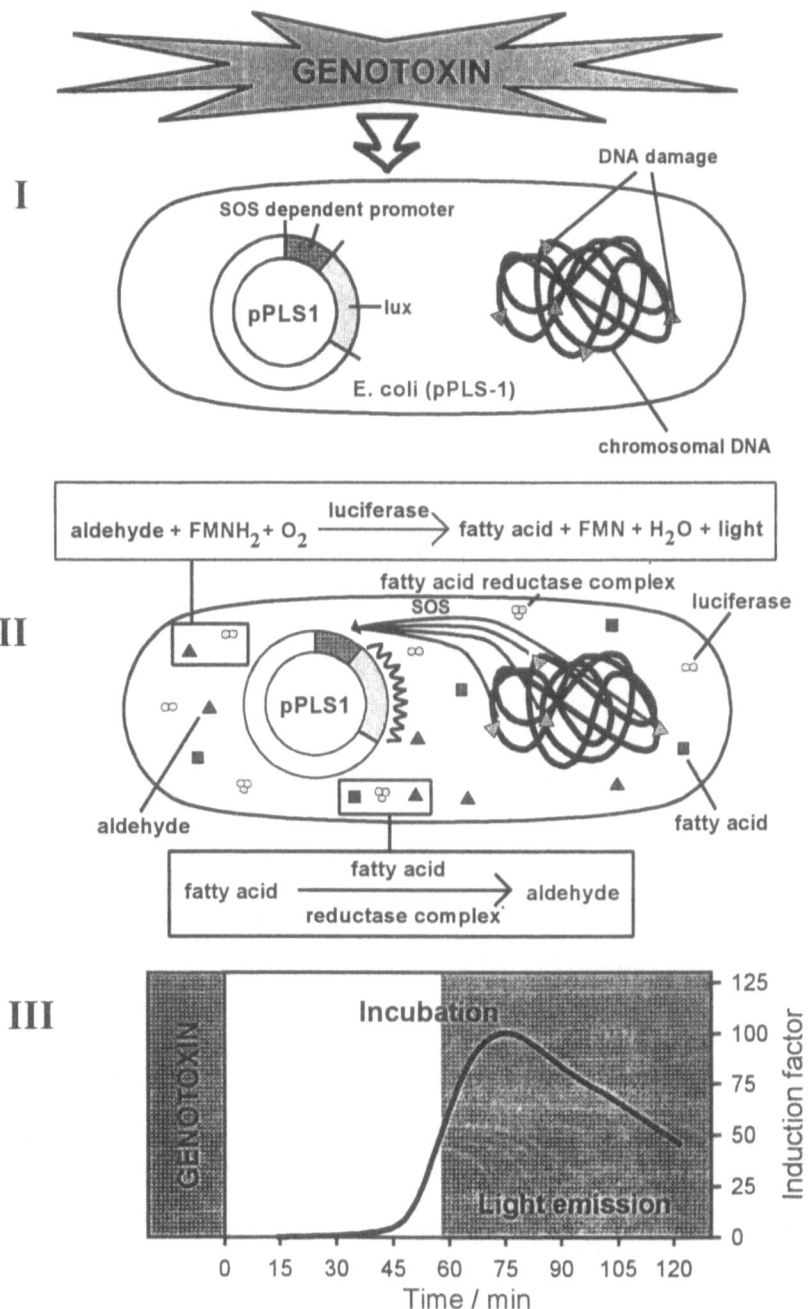

Figure 59: The steps of induction of the SOS *lux* system by a genotoxin

Heitzer et al. 1994). These biodetection systems stand out for a rapid induction of bioluminescence within approximately one hour in response to the environmental pollutant under concern.

5.1.2.3   The SOS *lux* test

We have constructed a bacterial biodetection system with specificity for genotoxins by combining the SOS system indicative for DNA damaging agents as receptor component with the bioluminescence system as a rapid reporter component. The recombinant *E. coli* C600 (pPLS1) carries a plasmid with the promoterless lux operon (*luxCDABFE*) of *Photobacterium leiognathi* under control of a strong SOS promoter that originates from part of the *cda* gene of the plasmid ColD (Frey et al. 1986). pPLS1 (DSM 10333) has all functions necessary for SOS inducible bioluminescence (Figure 59).

5.1.3   Detection of genotoxins

5.1.3.1   The factor of SOS *lux* induction

The factor of SOS induction F_i is equal to the relative light emission Lux_i/Lux_o divided by the relative OD_{560} values OD_x/OD_o:

$$F_i = (Lux_i \times OD_o)/(Lux_o \times OD_x)$$
(1)

with Lux_i = light emission of the culture treated with the genotoxin (mV),
    Lux_o = light emission of the culture without genotoxin (mV),
    OD_o = optical density of the untreated culture,
    OD_x = optical density of the treated sample.

This correction for cell concentration is necessary, because some genotoxins delay or impair cell growth and thereby influence the total light emission of the culture. F_i is plotted as a function of the dose of the genotoxin. The following criteria have been used to evaluate the degree of genotoxicity: (i) the lower limit of detection which is the dose at which the induction factor F_i reaches a value twice that of the background and (ii) the genotoxic potential which is the slope of the response curves at low doses. A substance is identified to be genotoxic, if at any of its concentrations F_i reaches a value of 2 or more. This or comparable criteria have been used in other SOS or mutagenictiy tests by several authors to evaluate the degree of genotoxicity (Ames et al. 1975, Quillardet and Hofnung 1993, Mersch-Sundermann et al. 1994). According to Mersch-Sundermann et al. (1992) a compound is considered "reactive genotoxic" if F_i exceeds 2.

5.1.3.2   Kinetics of SOS *lux* induction

In response to damage to their DNA, cells of the *E. coli* C600 (pPLS1) assay system induce the SOS system which includes a diverse set of physiological phenomena including SOS repair and SOS dependent bioluminescence. The kinetics of SOS induced bioluminescence are dependent on the bioavailability and dose of the geno-toxic agent as well as on a variety of cultivation conditions, such as concentration of the inoculum, composition of the growth medium, oxygen supply and growth tem-perature. Therefore, standardized procedures are required in order to secure high reproducibility of the assay. For the SOS *lux* bioassay exponentially growing cells of *E. coli* C600 (pPLS1) in L-medium (Miller 1972) are incubated with the genotoxin for 90 min at 37°C followed by 10 min of incubation at room temperature. Figure 60 shows the kinetics of SOS *lux* induction after incubation of the bioassay with 500 nM mitomycin C (MMC) at 37°C followed by 10 min of incubation at room temperature. This sequence of different temperatures is necessary, because the optimum temperature for SOS induction is at 37°C whereas bioluminescence reaches maximum

224

values at lower temperatures (about 20°C). Figure 60 shows that the bioluminescence signal increased from the baseline level continuously within the first 90 min of

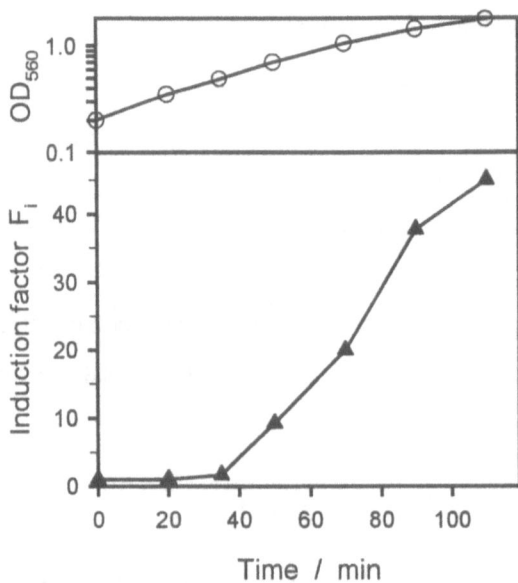

Figure 60: Kinetics of SOS *lux* induction by mitomycin C (MMC). The induction factor F_i and the cell density OD_{560} were determined after different incubation intervals at 37°C followed by 10 min incubation at room temperature

incubation. During this period the cell growth rate was nearly constant. Figure 60 shows also the growth curve of the untreated control, determined from the OD_{560} which was not significantly different from that of the MMC treated samples.

5.1.3.3 Sensitivity of SOS *lux* induction to MMC

MMC as a known genotoxin which predominantly induces DNA intrastrand cross-linking was used as an example for a chemical genotoxin. Figure 61 gives the dose response curve of the SOS *lux* test to MMC. In addition, the relative cell density is shown. The lower limit of detection, i.e. the dose at which F_i reaches a value of two, amounts to 4.3×10^{-9} M. The genotoxic potential, i.e. the slope of the dose response curve at low doses is 2.3×10^{8}/M.

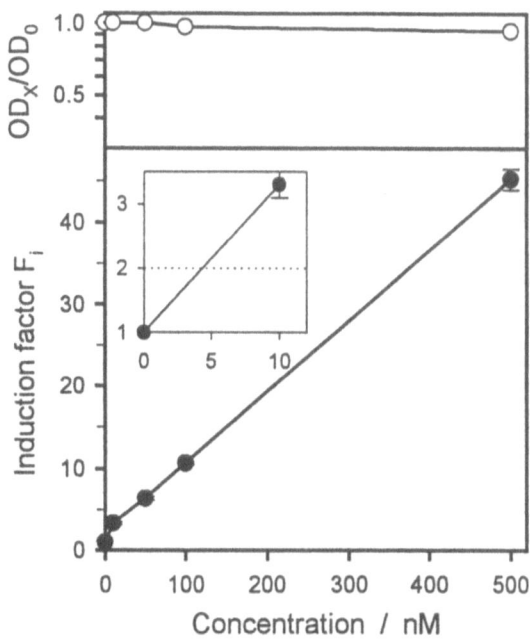

Figure 61: SOS *lux* induction curve for mitomycin C (MMC). The induction factor F_i and the relative cell density OD_x/OD_o were determined after 90 min incubation at 37°C followed by 10 min incubation at room temperature

5.1.3.4   Sensitivity of SOS *lux* induction to UV radiation

UV radiation at 254 nm specifically damages the DNA by inducing predominantly pyrimidine dimers. UV has been chosen as a physical genotoxin, because UV is a strong SOS inducer. Figure 62 shows the dose response curve of the SOS *lux* test to UV radiation.

The lower limit of detection amounts to 0.3 J/m², the genotoxic potential is 2.7 m²/J. It is interesting to note that at a dose of 20 J/m², the SOS *lux* induction curve still increases, whereas the relative cell density is significantly reduced. This shows the high genotoxic potency of UV radiation. In a future study, the SOS responsivity to UV-radiation will be correlated with the photoproducts induced in the DNA (in preparation).

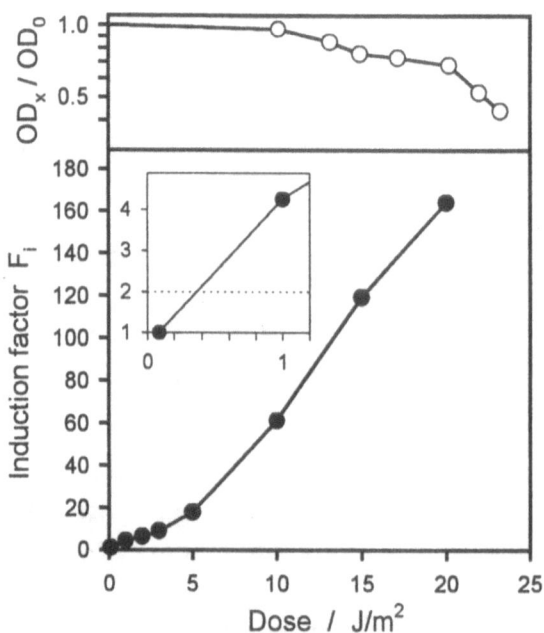

Figure 62: SOS *lux* induction curve for UV (254 nm) radiation. The induction factor F_i and the relative cell density OD_x/OD_o were determined after 120 min incubation at 30°C. All treatments were done under yellow light to prevent photoreactivation

5.1.4   Advantages of the SOS *lux* assay

5.1.4.1   Comparison with other genotoxicity tests

The SOS *lux* test is specific for the detection of genotoxins. Like other SOS based bacterial tests, such as the SOS chromotest (Quillardet et al. 1982) and the *umu* test (Oda et al. 1985) it makes use of the fact, that mutagenesis in *E. coli* cells exposed to radiation or chemical mutagens involves the SOS function which is induced by DNA damage, such as base changes or strand breaks. Therefore, SOS based systems have also been used for mutagenicity tests. By monitoring SOS gene expression, SOS bioassays specifically detect environmental genotoxic substances.

Table 25: Comparison of the sensitivity of the SOS *lux* test with that of other mutagenicity or genotoxicity tests

Genotoxin	Lower limit of detection[a] (M) for chemicals, (Gy) for γ rays			Revertants (nM^{-1}) or lower limit of detection[a] (Gy)
	SOS *lux* test	SOS chromotest[b] [c]	*umu* test[d]	Ames test[b] [e]
MMC	5.0×10^{-9}	1.7×10^{-8}	1.5×10^{-7}	1200
DMS	7.5×10^{-6}	6.7×10^{-6}	3.0×10^{-4}	0.1
MNNG	7.1×10^{-7}	5.0×10^{-7}	2.0×10^{-6}	44
γ rays	2.56	<5	n.d.	3 - 4

[a] Dose of the genotoxin to increase the SOS *lux* response by a factor of 2 over background.
[b] from Qillardet et al. (1982) for chemicals
[c] from Quillardet et al. (1989) for γ-rays
[d] from Nakamura et al. (1987)
[e] from Isildar and Bakale (1983) and Kozubek et al. (1989) for γ-rays
   n.d. = not determined

The SOS *lux* test was sensitive to the MMC at concentrations as low as the nanomolar level. Its sensitivity is comparable to that of other bioassays, such as the other mentioned SOS bioassays and the Ames test (Table 25).

5.1.4.2 Practical advantages of the SOS *lux* test

The SOS *lux* test as compared with other bioassays for mutagenicity and/or geno-toxicity has a number of practical advantages:

(1) Quick data acquisition. The test results are available within 1-2 h. The registration time is very short (≤1s). With this, the SOS *lux* test is faster than any other bioassay for mutagenicity and/or genotoxicity.

(2) Data acquisition without disrupture of the cells. The bioluminescent reporter technology allows *in vivo* analysis. The light signal is taken from the living cell,

the measurement is non-destructive for the test bacteria and can be repeated several times.

(3) <u>High precision of the measurements</u>. Repeated measurements of the SOS *lux* response within a few seconds substantially increase the precision of the data so that small deviations of F_i from 1 can be detected.

(4) <u>Continuous data acquisition.</u> The bioluminescence data of the test culture can be continuously monitored during the incubation period. This allows to record the whole time kinetics of SOS induction from the same culture. In principle, by use of a sensitive photodetector the light of one single cell can be detected.

(5) <u>Wide scope of application with the same tester strain</u>. Like other SOS bioassays, SOS *lux* test detects a wide range of genotoxins of different DNA damaging mechanisms, such as MMC, MNNG, NA, DMS, H_2O_2, by using the same tester strain (data not shown). In contrast, different strains are required for the detection of the mutagenic potency of these agents when using the Ames test. For example MMC is detected by *S. typhimurium* TA102 but not by strain TA100 (owing to defect in excision repair), H_2O_2 or ionizing radiation are effectively detected by strain TA102.

(6) <u>Possibility to use a wide variety of bacterial tester strains</u>. Depending on the type of genotoxin, the sensitivity of the SOS *lux* test can be largely increased by using different host strains for the pPLS1 plasmid. An example is UV radiation for which *E. coli* strains deficient in excision repair are much more sensitive than the wild type strain. Whereas the lower limit of detection was 0.3 J/m^2 when using the wild type *E. coli* C600 (pPLS1) strain, this value was lowered by one order of magnitude when using the *uvrA⁻* strain AB1886 which is deficient in excision repair (data not shown). Even higher sensitivity can be achieved by the use of the double mutant *umu⁻uvr⁻*. A similar increase in sensitivity is achieved for the SOS chromotest using an *E. coli* strain deficient in excision repair (Quillardet and Hofnung 1984). Furthermore, the permeability for some chemicals can be im-

proved using a strain carrying a *rfa* mutation. Recently, a strain possessing a high O-acetyltransferase activity was developed for the *umu* test, which is highly sensitive towards promutagenic aromatic amines (Oda et al. 1995).

(7) <u>Simultaneous test of cytotoxicity</u>. To prove the genotoxic effectiveness, a discrimination between genotoxic and cytotoxic potency of the test substance is needed. This has been achieved by simultaneous measurements of the cell concentration. If bioluminescence is not induced and the cell growth is comparable to that of the untreated control, the test substance is neither genotoxic nor cytotoxic. If however, bioluminescence and/or OD_{560} decrease during incubation, then the test suggests the agent in question to be cytotoxic.

5.1.5　Acknowledgement

This work was supported by COPERNICUS Grant CIPA CT - 94 0122 from the European Commission.

5.1.6　References

Ames, B.N., Lee, F.D., Durston, W.E. (1973): An improved bacterial test system for the detection and classification of mutagens and carcinogens. Proc. Nat. Acad. Sci. USA <u>70</u>, 782-786.

Ames, B.N., McCann, J., Yamasaki, E. (1975): Methods for detecting carcinogens and mutagens with the Salmonella/mammalian microsome mutagenicity test. Mut. Res. <u>31</u>, 347-364.

Burlage, R.S., Sayler, G.S., Larimer, F. (1990): Monitoring of naphthalene catabolism by bioluminescence with *nah-lux* transcriptional fusions. J. Bacteriol. <u>172</u>, 4749-4757.

Chatterjee, J., Meighen, E.A. (1995): Biotechnical applications of bacterial bioluminescence (*lux*) genes. Photochem. Photobiol. <u>62</u>, 641-650.

Corbisier, P., Nuyts, G., Ji, G., Mergeay, M., Silver, S. (1993): luxAB gene fusions with the arsenic and cadmium resistance operons of *Staphylococcus aureus* plasmid pI258. FEMS Microbiol. Lett. <u>110</u>, 2231-238.

Frey, J., Ghersa, P., Palacios, P.G., Belet, M. (1986): Physical and genetic analysis of the ColD plasmid. J. Bacteriol. <u>166</u>, 15-19.

Friedberg, E.C. (1985): DNA repair. Freeman, New York.

Hastings, J.W., Potrikas, C.J., Gupta, S.C., Kurfurst, M., Makemson, C. (1985): Biochemistry and physiology of bioluminescent bacteria. Adv. Microbial Physiol. <u>26</u>, 235-291.

Heitzer, A., Malachowski, K., Thonnard, J.E., Bienkowski, P.R., White, D.C., Sayler, G.S. (1994): Optical biosensor for environmental on-line monitoring of naphtha-lene and salicylate bioavailability with an immobilized bioluminescent catabolic reporter bacterium. Appl. Environ. Microbiol. <u>60</u>, 1487-1494.

Isildar, M., Bakale, G. (1984): Radiation-induced mutagenicity and lethality in Ames tester strains of *Salmonella*. Rad. Res. <u>100</u>, 396-411.

ISO CD 13829 (1996): Water quality- determination of the genotoxicity of water and waste water using the „umutest", ISO/TC 147/SC5 Biological Methods.

Kozubek, S., Krasavin, E.A., Amiratyev, K.G., Tokarova, B., Soska, J., Drásil, V., Bonev, M. (1989): The induction of revertants by heavy particles and γ-rays in *Salmonella* tester strains. Mut. Res. <u>210</u>, 221-226.

Lewis, R.J. Sr. (1991): Carcinogenically active chemicals: a reference guide. Van Nostrand Reinhold, New York.

Maron, D.M., Ames, B.N. (1983): Revised methods for the *Salmonella* mutagenicity test. Mut. Res. <u>113</u>, 173-215.

McDaniels, A.E., Reyes, A.L., Wymer, L.J., Rankin, C.C., Stelma, G.N. (1990): Comparison of the *Salmonella* (Ames) test, *umu* tests, and the SOS chromotests for detecting genotoxins. Environ. Mol. Mutagen. <u>16</u>, 204-215.

Mersch-Sundermann, V., Mochayedi, S., Kevekordes, S. (1992): Genotoxicity of polycyclic aromatic hydrocarbons in *Escherichia coli* PQ37. Mut. Res. 278, 1-9.

Mersch-Sundermann, V., Schneider, U., Klopman, G., Rosenkranz, H.S. (1994): SOS induction in Escherichia coli and *Salmonella* mutagenicity: a comparison using 330 compounds. Mutagenesis 9, 205-224.

Miller, J. (1972): Experiments in Molecular Genetics. Cold Spring Harbor Laboratory Press, Cold Spring Harbor, NY.

Quillardet, P., Hofnung, M. (1984): Induction by UV light of the SOS function sfiA in Escherichia coli strains deficient or proficient in excision repair. J. Bacteriol. 157, 35-38.

Nakamura S., Oda, Y., Shimida, T., Oki, I., Sugimoto, K. (1987): SOS-inducing activity of chemical carcinogens and mutagens in *Salmonella typhimurium* TA1535/pSK1002: examination with 151 chemicals. Mut. Res. 192, 239-246.

Oda, Y., Nakamura, S., Oki, I., Kato, T., Shinagawa, H. (1985): Evaluation of the new system (umu-test) for the detection of environmental mutagens and carcinogens. Mut. Res. 147, 219-229.

Oda, Y., Yamazaki, H., Watanabe, M., Nohmi, T., Shimada, T. (1995): Development of high sensitive umu test system: rapid detection of genotoxicity of promutagenic aromatic amines by *Salmonella typhimurium* strain NM2009 possessing high O-acetyltransferase activity. Mut. Res. 334, 145-156.

Quillardet, P., Huisman, O., Ari, R.D., Hofnung, M. (1982): SOS chromotest, a direct assay of induction of an SOS function in *Escherichia coli* K12 to measure geno-toxicity. Proc. Natl. Acad. Sci. USA 79, 5971-5975.

Quillardet, P., Hofnung, M. (1984): Induction by UV light of the SOS function sfiA in Escherichia coli strains deficient or proficient in excision repair. J. Bacteriol. 157, 35-38.

Quillardet, P., Frelat, G., Nguyen, V.D., Hofnung, M. (1989): Detection of ionizing radiation with the SOS chromotest, a bacterial short-term test for genotoxic agents. Mutation Res. 216, 251-257.

Quillardet, P., Hofnung, M. (1993): The SOS chromotest: a review. Mut. Res. 297, 235-279.

Reifferscheid, G., Heil, J., Oda, Y., Zahn, R.K. (1991): A microplate version of the SOS/*umu*-test for rapid detection of genotoxic potential of environmental samples. Mut. Res. 253, 215-222.

Sassanfar, M., Roberts, J.W. (1990): Nature of SOS-inducing signal in *E. coli*. The involvement of DNA replication. J. Molec. Biol. 212, 79-96.

Scheller, F., Schmid, R.D. (1991): Biosensors: fundamentals, technologies and applications. VCH Verlagsgesellschaft, Weinheim.

Selfinova, O., Burlage, R., Barkay, T. (1993): Bioluminescent sensors for detection of bioavailable Hg(II) in the environment. Appl. Environ. Microbiol. 59, 3083-3090.

Stewart, G.S.A.B., Denyer, S.P., Lewington, J. (1991): Microbiology illuminated: gene engineering and bioluminescence. Trends Food Sci. Technol. Today 2, 19-22.

Tescione L., Belfort, G. (1993): Construction and evaluation of a metal ion biosensor. Biotech. Bioengin. 42, 945-952.

White, P.A., Rasmussen, J.B. (1996): SOS chromotest results in a broader context: empirical relationships between genotoxic potency, mutagenic potency, and carcinogenic potency. Environ. Mol. Mutagen. 27, 270-305.

Witkin, E.M. (1976): Ultraviolet mutagenesis and inducible DNA repair in *Escherichia coli*. Bacteriol. Rev. 40, 869-907.

6   Field Experiments

6.1 Potential and Capabilities of Biosensors for the Assessment of Environmental Pollutants

Karl Cammann, Gabriele Chemnitius, Markus Meusel and Christiane Zaborosch

ICB, Institut für Chemo- und Biosensorik e. V., Mendelstr. 7, D-48149 Münster, Germany

Abstract. By exploiting the unique features of bio-recognition systems biosensors have the potential to complement conventional analytical methods. Especially when offering new or improved capabilities over existing technologies such as the estimation of biological effects, sum-values, or the potential for analysis in complex and divers natural matrices these devices become very useful for several environmental applications. In this context immunosensors for herbicide and PAH-determination, and enzyme sensors for the determination of phosphate and phenol are presented. Moreover, microbial sensors for the estimation of biological oxygen demand and the assessment of environmental pollutants are described.

6.1.1   Introduction

According to IUPAC definition a biosensor is a self-contained integrated system which is capable of providing specific quantitative or semi-quantitative analytical information using a biological recognition element which is directly spatially coupled to a transducer element. The two most important properties of any proposed biosensor are its specificity and its sensitivity towards the target analyte(s). The specificity is principally governed by the properties of the biological component. The sensitivity of the integrated device, however, is dependent on both the biological component and the transducer because there must be a significant biomolecule-analyte interaction and a high efficiency of subsequent detection of this molecular recognition process by the transducer.

The relatively slow progress of biosensor technology from inception to fully functional commercial devices for these applications is a reflection of both technology-related and market factors. At the moment a wider acceptance of biosensors as reliable analytical tools is impaired by their useful active life and stability. Improvements in the stability and retention of biochemical activity in *in vitro* environments can thus be considered vital to the success of these devices.

In contrast to conventional analytical devices biosensors have potential for continuous and *in situ* applications, and they are suitable for a wide variety of matrices. Their potential for environmental applications lies in the ability to measure the interactions of pollutants with the biological system, i.e. an antibody, an enzyme or a whole microorganism. Especially the possibility to assess biological *effects* contrasts with the abilities of routine analytical methods. The oxygen consumption of whole microorganisms, for example, can be correlated to the presence of organic pollutants, or the cross-reactivity of an antibody molecule can be exploited by developing immunosensors for the estimation of overall sum-values. Only by offering new capabilities or improvements over existing and established methods biosensors will be commercially successful.

6.1.2 Results and discussions

6.1.2.1 Immunosensor systems

It is one of the major benefits of immunosensors that antibodies can be obtained to any kind of chemical structure (Marco et al. 1995). As the specificity of the sensor is almost entirely governed by the antibody molecule the system can be easily adapted to another analyte just by changing the corresponding antibody. If the sensor system should be constructed for highly selective single-compound measurements an antibody molecule with almost no crossreactivities towards structurally similar compounds can be selected. Using polyclonal antibodies or a monoclonal antibody with broad cross-reactivities as the recognition element, however, the sensor can be optimised for the estimation of an overall sum-value below or above a certain threshold concentration and, thus, would be the ideal tool for screening purposes.

With regard to this aspect we developed flow-injection immunoanalysis (FIIA)-systems for the selective determination of the herbicide 2,4-dichlorophenoxyacetic acid (2,4-D) in water and for the estimation of an overall PAH-(polycyclic aromatic hydrocarbons) concentration in soil.

FIIA-system for herbicide determination

The herbicide 2,4-D is a weedkiller widely applied in agriculture. The herbicide may contaminate not only ground and surface water, but drinking water as well. Thus the highest permissible concentration in drinking water is 0.1 µg/l 2,4-D.

The FIIA set-up comprises an autosampler, a selector, a peristaltic pump, the immunoreactor, an amperometric detector and a potentiostat (Chemnitius et al. 1996). The whole system is computer-controlled and almost entirely automated. The most important part of the FIIA system is the immunoreactor. In previous experiments a fused silica capillary used for GC-analysis was applied. This reactor, however, was very fragile and difficult to handle. In parallel a chip-immunoreactor (10 x 20 mm) manufactured in microsystem technology was developed at the Fraunhofer Institute for solid phase technology, IFT, in Munich (Woias et al. 1996). Current experiments aim at the replacement of this reactor by simple and cheap reactors manufactured from polymers and glass.

The herbicide 2,4-D was determined in a titration assay using highly specific monoclonal anti-2,4-D antibodies as described earlier (Chemnitius et al. 1996, Trau et al. 1997). For electrochemical detection alkaline phosphatase was used as enzyme label together with p-aminophenyl phosphate as substrate (Tang et al. 1988). Enzymatically generated p-aminophenol was subsequently detected at a carbon working electrode (+150 mV, vs. Ag/AgCl). After each measurement the immunoreactor was regenerated by running a low pH buffer through the column, stripping off bound antibodies. A decrease in pH value is most commonly used to weaken the specific antibody-antigen interaction and to remove bound immuno-complexes. In this particular work the immunoreactor was regenerated by applying a glycine/HCl buffer pH 2.7. A comparison of different immunoreactors (Figure 63) demonstrates that there is no significant difference between the calibration curves obtained with either the capillary-reactor or the chip-reactor.

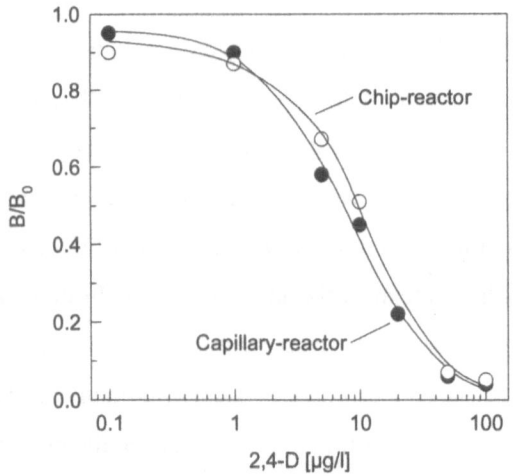

Figure 63: Comparison of different immunoreactors. Standardised calibration curves for 2,4-D

Differences in signal intensities of several hundred nA (between analyte concentrations from 0 to 100 µg/l 2,4-D) offered measurements with a high sensitivity.

The detection limit is in the range of 1 µg/l 2,4-D. It has to be emphasised that the sensor system avoids time consuming sample pre-treatments, e.g. solid phase extraction with organic solvents, and that the analyte is detected within a sample volume of 2 ml only (in contrast to HPLC-analysis which requires 2 l of sample).

Although an even lower detection limit has to be envisaged for practical herbicide determination the major immanent advantages of this biosensor technology, such as the short analysis time of 10 min only, the continuous monitoring capability (*via* regeneration of the immunoreactor) and the relinquishment of organic solvents or additional sample pre-treatment become obvious. The critical step in the immunochemical measurement, however, is the stability of the immunoreactor. Due to a loss of „activity" of the immobilised biocomponents the lifetime of the reactor is limited to approximately 100 measuring cycles. New approaches, based for example on the construction of new assay formats, however, will help to speed up significant improvements in sensor lifetime and thus will help to expedite commercialisation of these kind of sensors.

Immunosensor for PAH determination in soil

Another challenge in environmental applications is the development of field screening and monitoring methods. These methods are for example particularly well suited for the rapid initial characterisation of hazardous waste sites whereas remediation depends on the presence of a threshold concentration of a certain pollutant. Among the variety of instruments and devices especially field test kits are relatively mature, resulting in a number of commercial products (Rogers and Williams 1995).

For the estimation of an overall soil contamination with polycyclic aromatic hydrocarbons (PAH) a FIIA-system similar to that described for herbicide determination was developed. PAH are hazardous substances due to their deleterious effects on human health and the ecosystem. In contrast to the sensor system for 2,4-D determination the FIIA-system was optimised for the estimation of a sum-value, by

238

using polyclonal antibodies raised against phenanthrene and by exploiting the broad cross-reactivities of the antibodies towards other PAH. Following the sensing principle described above phenanthrene was coupled to a carrier protein and immobilised on the inner surface of the immunoreactor. For sample preparation soil samples were extracted with methanol by simple shaking followed by a dilution step with buffer. When applied to the immunoreactor PAH in the sample compete with the immobilised phenanthrene for binding to the anti-phenanthrene antibodies.

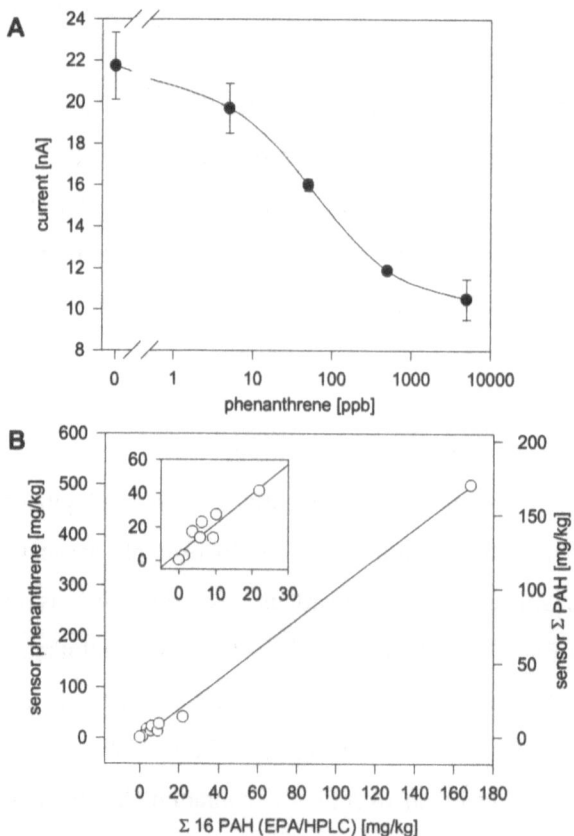

Figure 64: Determination of total PAH in methanolic extracts of soil. A, calibration curve for phenanthrene. B, correlation study. Comparison of HPLC measurements (16 EPA-PAH) and sensor response as „sensor phenanthrene". The y-axis on the right side shows the sensor „total PAH value" calculated from the „sensor phenanthrene" values and the empirical correlation factor of 0.3

The sensor was calibrated with phenanthrene standards (Figure 64A). This hazardous substance could be determined within the range of 20 to 200 ppb. To estimate the precision of the sensor various soil samples were analysed with the immunochemical method and HPLC (Figure 64B). For HPLC-analysis the concentration of 16 PAH (according to EPA-method 8310) was determined in methanolic extracts. Figure 64B shows a good correlation between the sum of the 16 EPA-PAH (x-axis) and the sum-value of the immunosensor (left y-axis).

Due to the cross-reactivities of the antibody and the presence of not only the 16 EPA-PAH but a vast number of structurally related compounds and PAH-metabolites the sensor signal is significant higher than the sum-value determined with HPLC.

For the estimation of an unknown PAH concentration the sensor value has to be multiplied with the empirical correlation factor of 0.33 (right y-axes). A prerequisite for this procedure is a constant ratio of phenanthrene to the other PAH in the sample which was confirmed with HPLC analysis.

The value of immunochemical methods has already been demonstrated for many field applications. The complex and diverse nature of natural matrices such as waste water or soil, however, poses serious problems. Especially the sample preparation problem is a significant challenge to be addressed in the future. Combining the selectivity and sensitivity of conventional immunoassay techniques with the capability of sensor technology for quantitative analysis will broaden environmental biosensor applications in the future.

6.1.2.2 Enzyme sensor systems

Several environmental hazardous substances can specifically be detected by the use of electrochemical enzyme sensors. These sensors comprise one or more enzymes combined in enzyme cascades or parallel enzyme reactions which are responsible for the selective analyte recognition. Compared to many well known and established reference methods enzyme sensors are at least superior with regard to environmental compatibility as no additional chemicals have to be added. A special feature of these sensors is the possibility of chemical or biochemical signal amplification which will be explained in detail in the following examples. Both, the use of enzymes with broad substrate or inhibitory spectra and the combination of several different enzymes to enzyme arrays will result in sensitive detection of whole groups of substances. Furthermore enzyme sensors can be integrated into automated and portable flow systems for *on-line* analysis.

Determination of inorganic phosphate

Inorganic phosphate plays an important role in eutrophication of lakes and rivers. A sensitive enzyme sensor for determination of inorganic phosphate was developed by use of the four enzyme sequence shown in Figure 65. The sequence consists of maltose phosphorylase, phosphatase, mutarotase and glucose oxidase co-immobilised onto a platinum working electrode (Conrath et al. 1995). The key enzyme maltose phosphorylase was purified at ICB from *Lactobacillus brevis* in a three step process.

In the presence of inorganic phosphate maltose is specifically hydrolysed to glucose-1-phosphate and α-D-glucose by maltose phosphorylase. During phosphate measurements the concentration of the cosubstrate maltose has to be kept constant.

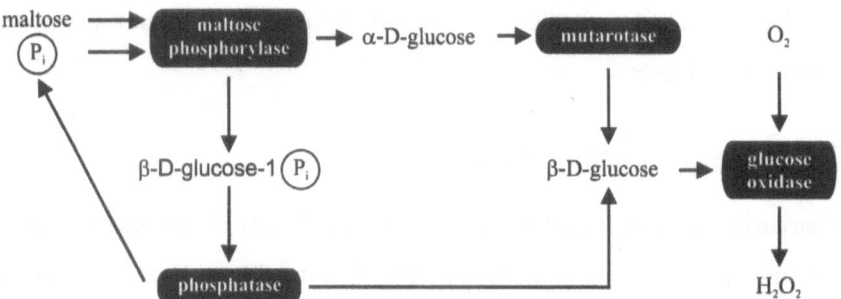

Figure 65: Four enzyme sequence involving signal amplification for the detection of inorganic phosphate

The sensor signal is dependent on the mutarotation rate of α-D-glucose to β-D-glucose. Thus coimmobilisation of mutarotase raises sensor signal height and reduces the sensor response time. The indicating enzyme glucose oxidase converts β-D-glucose. The hydrogen peroxide produced is measured at a platinum working electrode at + 600 mV vs. Ag/AgCl. Further signal amplification was achieved by co-immobilisation of phosphatase hydrolysing the phosphomonoester to glucose and phosphate. In this way phosphate is recycled entering the detection cycle several times.

During previous experiments phosphate determination was performed using a macro Pt-electrode in a batch system. Due to the amplification system a very low detection limit of 10 nM inorganic phosphate could be achieved. Out of several heavy metal ions tested only Cd^{2+} and Zn^{2+} showed inhibitory effects on the enzyme sequence. In contrast to chemical sensors for the determination of phosphate based on potentiometric anionselective electrodes using Ag/Ag_3PO_4 or different bicarbonates showing high cross reactivity especially to halides even in low concentrations, the sensitive enzyme sequence of the presented phosphate sensor is not influenced by halide concentrations below 0.2 M.

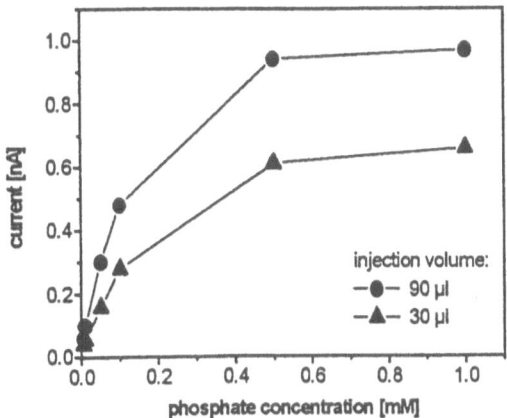

Figure 66:   Calibration curves obtained with the four-enzyme sequence immobilised onto a planar microelectrode array. Maltose concentration in the carrier was 0.2 mM. Different sample volumes were injected

For automated sample analysis a hybrid micro flow injection analysis system has been constructed. In this FIA system different microelectrode arrays fabricated in thin film technology were used as working electrodes in a thermostated flow through cell. Sensors were completed by *on-chip* counter and reference electrodes. For carrier propelling conventional roller pumps were used and magnetic valves were applied for sample introduction. The system was very flexible with respect to sample volume as the switching of the different valves can form different injection loops. Therefore a calibration curve could be obtained using only one single sample concentration which was injected in several volumes (Figure 66). All fluidic components were located on the top of the apparatus for ease of access. Within this automated FIA-system two- and threedimensionally structured microelectrode arrays were used as transducers for biosensor development. With the two-dimensional array a lower detection limit of 5 µM was achieved. Using the three-dimensional array the detection limit was 1 µM. The working stability of the system was more than a week in continuous use with sample injection every 5 minutes. Moreover the system has successfully been applied to the determination of inorganic phosphate in surface water. Comparison of the results with those obtained by a reference laboratory with the molybdenum blue method showed an excellent correlation.

Determination of phenols

The determination of phenolic compounds was accomplished by use of tyrosinase as specific indicating enzyme. Tyrosinase is known for two different enzymatic activities. On the one hand it catalyses the hydroxylation of phenols to *o*-phenols and on the other hand it catalyses their subsequent oxidation to *o*-quinones. A kind of sum parameter similar to the so called „phenol index" is provided by the biosensor as not only phenol itself may serve as a substrate of tyrosinase but also various substituted phenols as well. There are two possibilities for signal generation. Without the addition of an electrochemically active mediator, o-quinone, the enzymatic reaction product can be directly reduced at the working electrode (Skládal 1991). But as this direct reduction also generates side products leading to electrode fouling and thus weak sensor stability use of a mediator is recommended (Kotte et al. 1995). In this way much more sensitive sensors can be obtained because of employing a mediator also results in signal amplification by recyclisation of enzymatically formed o-quinones and efficient reduction of the mediator at the working electrode (Figure 67).

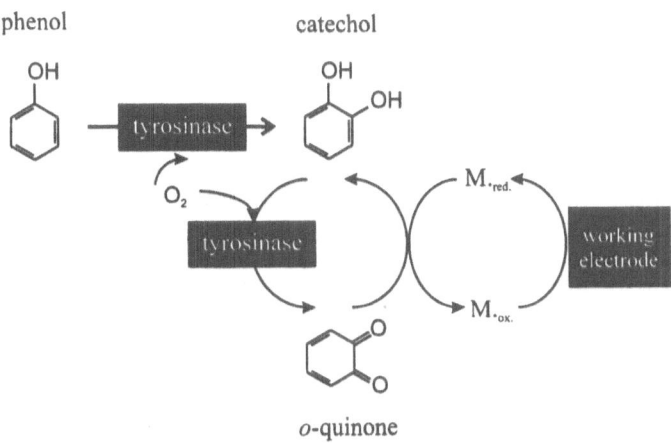

Figure 67: Reaction scheme of the determination of phenols by a tyrosine based biosensor. The enzymatic reaction as well as the mediator catalysed recycling of catechol and the signal generation at the working electrode are shown

The development of miniaturised sensor systems for water monitoring favours miniaturised enzyme sensors fabricated in silicon technology which can easily be integrated into such systems.

Therefore, immobilisation procedures like electropolymerisation are desirable allowing spatially controllable deposition of enzymes (Trojanowicz et al. 1995). Figure 68 shows a calibration curve of a sensor with tyrosinase immobilised by entrapment in an electropolymerised mediator modified pyrrole.

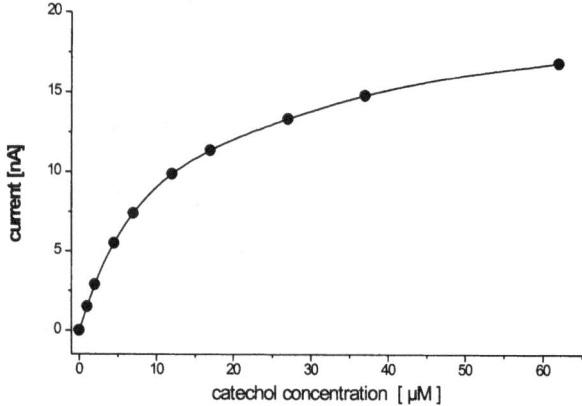

Figure 68: Catechol determination in a batch experiment. Tyrosinase was immobilised by entrapment in a gallocyanine modified polypyrrole electropolymer. 1 mm diameter Pt working electrode, working electrode potential was -50 mV vs. Ag/AgCl/3M KCl

6.1.2.3 Microbial sensors

Microbial sensors using immobilised cells as recognition elements in combination with a physical transducer show some advantages over biosensors based on proteins such as antibodies or enzymes. In general, they are less sensitive to inhibition and more tolerant of suboptimal pH and temperature and often have a higher stability. Because microorganisms perform multistep transformations microbial sensors allow the sensitive determination of a wide spectrum of substances in various fields, especially in pollution control. The huge diversity of microorganisms with a biodegradation capacity of natural substances and xenobiotics is an inexhaustible reservoir for many applications of biosensors.

Determination of PAH in aqueous samples

Microbial sensors for the determination of toxic compounds applicable for an environmental monitoring were developed for example for the determination of chlorophenols (Riedel et al. 1995), benzene (Tan et al. 1994) and phenol (Rainina et al. 1996). In the following a microbial sensor system for the detection of naphthalene in aqueous samples is presented (König et al. 1996). The sensor can be used for the determination of bioavailable and water soluble PAH whereas the PAH-FIIA-system described above was tailored for the assessment of extractable water insoluble compounds.

As outlined above polycyclic aromatic hydrocarbons represent the largest group of cancerogenic compounds present in the environment. They are ubiquitous and especially found in contaminated areas of former gasworks and cooking plants. Conventional analysis of PAH by high performance liquid chromatography has a low detection limit and enables the discrimination between single compounds, but is expensive, time consuming and requires large quantities of pollutant organic solvents for sample preparation. As biosensors based on microorganisms do not need sample preparation, can be produced at reasonable cost compared to standard analytical equipment and allow a higher sample throughput, they may be a useful tool for the rapid screening of a large number of samples. The biosensors developed are based on microorganisms specialised in degradation of PAH. As microbial degradation of PAH is performed under O_2 consumption, an amperometric transducer principle is applicable. In the presence of assimilable analytes the oxygen consumption increases proportionally to the concentration of these substances. Due to its high water solubility, naphthalene is often found in aqueous extracts of contaminated soils as a main component and was therefore chosen as leading analyte.

Two bacterial strains, *Pseudomonas fluorescens* WW4 and *Sphingomonas* sp. B1 which are able to grow on mineral medium with PAH as sole source of carbon were used for the sensors. The microbial cells were immobilised within a novel elastic and non-toxic polyurethane based hydrogel (Vorlop et al. 1992). The respiratory activity of the immobilised microorganisms was measured with a miniaturised Clark-type oxygen electrode. Both strains - *P. fluorescens* and *Sphingomonas* sp. - showed identical performance in the determination of naphthalene. The sensors reached detection limits of 0.01 mg/l which is below threshold values for the remediation of ground water stated in the List of the Netherlands (0.07 mg/l). The linear range (0.01 - 3.0 mg/l) of our biosensors corresponds to the concentrations found in aqueous extracts of contaminated soil (up to 1 mg/l). The sensors showed a response time of 2 min and a repeatability of ± 5 % and allowed a sample throughput of 6 per hour. The sensors reached an operational lifetime of up to 20 days.

The sensitivity of the sensors for naphthalene as leading analyte was in most cases more than four times higher than for various other substrates such as carbo-hydrates, organic acids, amino acids, and alcohols. The maximum signals when meas-uring these substrates were generally much lower than with naphthalene or

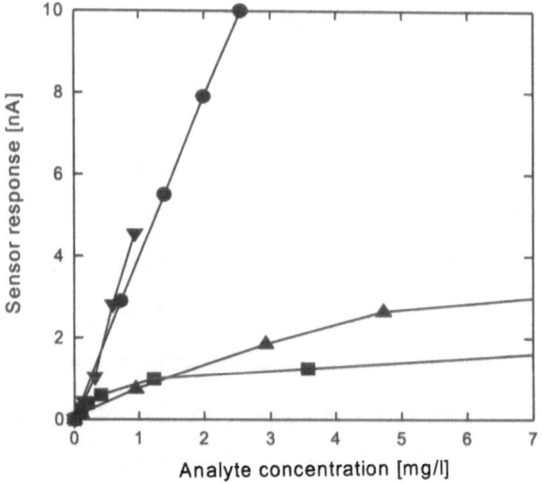

Figure 69: Calibration curves of the *Sphingomonas* sp. containing sensor for naphthalene (●), phenanthrene (▼), acetate (▲) and benzoate (■). Calibration was continued up to 22.4 mg/l acetate and 31.6 mg/l benzoate, data points not shown

phenanthrene as another compound of the PAH family (Figure 69). Moreover, it was possible to detect naphthalene with unchanged sensitivity even when another substrate was present at saturation concentration.

Estimation of biochemical oxygen demand

The biosensor with most widespread application in controlling waste water pollution is the one for the detection of biochemical oxygen demand (BOD). This widely used parameter indicates the content of biodegradable organic compounds in waste water and surface waters. The conventional BOD_5 method uses undefined bacterial mixed cultures, shows a high standard error of approx. 20 %, requires five days and is therefore not practicable for process control of waste water treatment. BOD can be estimated more rapidly by using the amperometric microbial sensor device as described (Karube et al. 1977, Riedel 1993). In the meantime several sensor BOD-systems were commercialised by German and Japanese companies. However, heavy metal ions which can be present in waste water cause inhibitory or toxic effects on the microorganisms in these sensors. This general problem of all commercially available BOD sensors can be avoided by using heavy metal resistant strains.

We developed a sensor on the basis of the heavy metal resistant strain *Alcaligenes eutrophus* KT02 (Slama et al. 1995). This organism isolated from a sewage plant (Timotius et al. 1987) carries plasmids encoding heavy metal resistance to 40 mM $NiCl_2$, 20 mM $CoCl_2$, 10 mM $ZnCl_2$ and 1 mM $CdCl_2$ (Schmidt et al. 1991). The cells were entrapped within a polyurethane-hydrogel between two membranes in front of a miniaturised Clark oxygen electrode. The sensor was integrated in a flow through system with alternating flow of buffer and sample. The excellent repeatability of approx. 2 % and the high sampling rate of 8 analyses per hour of this sensor even allow for continuous monitoring, e. g. in waste water treatment. The sensor shows a response time of 80 s only and has an operational lifetime of more than 30 days.

The tolerance of the sensor towards heavy metal ions was examined by analysing a standard solution in the presence and absence of heavy metal ions. The sensor was repeatedly exposed for 30 s to the standard containing additionally 4 mM heavy metal ions (Figure 70, left graph). For comparison all sensor responses in the presence of heavy metal ions were normalised to the standard response without metal ions recorded before each metal measuring period. Obviously, the heavy metals nickel, copper and zinc are tolerated over a period of 10 h and cadmium for 4 h with a sampling rate of 8 analyses per hour. After this treatment the recovery of the sensor was examined in the absence of heavy metal ions. The right graph demonstrates that even after 10 h exposure to cadmium the sensor recovers to its original performance. In contrast, conventional sensors equipped with normal microorganisms are in most cases irreversibly damaged after only 3 measurements. Our experiments clearly show that a sensor based on a heavy metal resistant strain is capable of measuring the BOD of waste water contaminated by such concentrations of heavy metal ions.

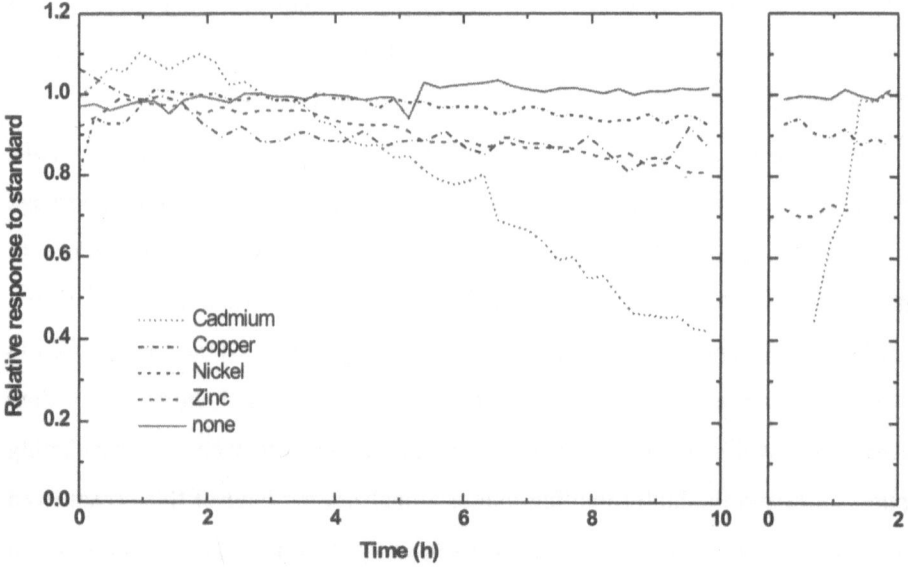

Figure 70: Influence of heavy metal ions on the sensor using *Alcaligenes eutrophus* KT02, standard: 2 mg/l glycerol without addition of metal ions (left graph: measurements in presence of 4 mM heavy metal ions; right graph: subsequent measurements without heavy metal ions)

Simultaneous multicomponent analysis with a single dynamic microbial sensor

Conventional biosensors use a biological component which is as selective as possible for a single analyte. Biosensors on the basis of immobilised whole cells represent an interesting exception. Each microorganism is provided with a multitude of different enzymes which results in a wide range of biosensor applications. On the one hand microbial sensors offer the possibility to determine complex parameters, e.g. biochemical oxygen demand as shown above, or substance families as PAH. On the other hand the determination of single analytes is feasible if suitable membranes for specific separation are available.

Recently, a procedure was developed which allows a much more flexible application of microbial sensors (Slama et al. 1996). Processes like transport and metabolism of the microorganisms cause an individual time dependent response in oxygen consumption for different pure organic analytes. Chemometrical data analysis revealed that the individual responses of single analytes in mixtures are additive and linear. Therefore simultaneous multicomponent analysis of organic mixtures with a limited number of known components is realised using only a single dynamic microbial sensor in combination with a partial least squares regression method. For multicomponent analysis and monitoring in biotechnological processes like enzymatic or microbial conversions the described dynamic microbial sensor offers an alternative to expensive analytical equipment.

6.1.3   Conclusions

As demonstrated the unique features of biosensors offer various interesting analytical possibilities. Although competing with established conventional analytical methods sensors or field tests become especially useful analytical tools when offering new or improved capabilities such as the estimation of biological effects, sum-values, etc. The relatively slow progress of biosensor technology from inception to fully-functional commercial devices in not only a reflection of market factors but technology-related factors as well. Thus it is not only essential to find niche applications but also to improve the stability and reliability of these devices in real environments.

6.1.4　References

Chemnitius, G., Meusel, M., Zaborosch, C., Knoll, M., Spener, F., Cammann, K. (1996): Highly sensitive electrochemical biosensors for water monitoring. Food Technol. Biotechnol. 34, 23-29.

Conrath, N., Gründig, B., Hüwel, S., Cammann, K. (1995): A novel enzyme sensor for the determination of inorganic phosphate. Anal. Chim. Acta 309, 47-52.

Karube, I., Matsunaga, T., Mitsuda, S., Suzuki, S. (1977): Microbial electrode BOD sensors. Biotech. Bioeng. 19, 1535-1547.

König, A., Zaborosch, C., Muscat, A., Vorlop, K.-D., Spener, F. (1996): Microbial sensors for naphthalene using *Sphingomonas* sp. B1 or *Pseudomonas fluorescens* WW4. Appl. Microbiol. Biotechnol. 45, 844-850.

Kotte, H., Gründig, B., Vorlop, K., Strehlitz, B., Stottmeister, U. (1995): Methylphenazonium-modified enzyme sensor based on polymer thick films for subnanomolar detection of phenols. Anal. Chem. 67, 65-70.

Marco, M.-P., Gee, S., Hammock, B.D. (1995): Immunochemical techniques for environmental analysis. I. Immunosensors. Trends Anal. Chem. 14/7, 341-350.

Rainina, E., Badalian, I., Ignatov, O., Fedorov, A. Simonian, A., Varfolomeyev, S. (1996): Cell biosensor for detection of phenol in aqueous solutions. Appl. Biochem. Biotechnol. 56, 117-127.

Riedel, K. (1993): Rapid measurement of Biochemical Oxygen Demand by using microbial sensors - a review. Vom Wasser 81, 243-256.

Riedel, K., Beyersdorf-Radeck, B., Neumann, B., Scheller, F. (1995): Microbial sensors for determination of aromatics and their chloroderivatives. Part III: Determination of chlorinated phenols using a biosensor containing *Trichosporon beigelii (cutaneum)*. Appl. Microbiol. Biotechnol. 38, 556-559.

Rogers, K., Williams, L.R. (1995): Biosensors for environmental monitoring: a regulatory perspective. Trends Anal. Chem. 14, 289-294.

Schmidt, T., Stoppel, R.-D., Schlegel, H.G. (1991): High-level nickel resistance in *Alcaligenes xylosoxydans* 31A and *Alcaligenes eutrophus* KT02. Appl. Environ. Microbiol. 57, 3301-3309.

Skládal, P. (1991): Mushroom tyrosinase-modified carbon paste electrode as an amperometric biosensor for phenols. Collect. Czech. Chem. Commun. 56, 1427-1433.

Slama, M., Zaborosch, C., Spener, F. (1995): Microbial sensor for rapid estimation of the Biochemical Oxygen Demand (BOD) in presence of heavy metal ions, p 171-174. In: Int. Conference - Heavy Metals in the environment, Vol. 2. (Wilken, R.D., Förstner, U., Knöchel, A., eds.). CEP Consultants Ltd., Edinburgh.

Slama, M., Zaborosch, C., Wienke, D., Spener, F. (1996): Simultaneous mixture analysis using a dynamic microbial sensor combined with chemometrics. Anal. Chem. 68, 3845-3850.

Tan, H.-M., Cheong, S.P., Tan, T.-C. (1994): An amperometric benzene sensor using whole cell *Pseudomonas putida* ML2. Biosens. Bioelectron. 9, 1-8.

Tang, H.T., Lunte, C.E., Halsall, H.B., Heineman, W.R. (1988): p-Aminophenyl phosphate: an improved substrate for electrochemical enzyme immunoassay. Anal. Chim. Acta 214, 187.

Timotius, K., Schlegel, H.G. (1987): Aus Abwässern isolierte nickel-resistente Bakterien. Nachr. der Akademie der Wissenschaften, Göttingen, II. Math.-Pysik. Klasse 3, 15-23.

Trau, D., Theuerl, T., Wilmer, M., Meusel, M., Spener, F. (1997): Development of an amperometric flow injection immunoanalysis-system for the determination of the herbicide 2,4-dichlorophenoxyacetic acid in water. Biosens. & Bioelectron. in press.

Trojanowicz, M., Geschke, O., Krawczynski vel Krawczyk, T., Cammann, K. (1995): Biosensors based on oxidases immobilised in various conduction polymers. Sens. Actuators. B 28, 191-199.

Vorlop, K.-D., Muscat, A., Beyersdorf, J. (1992): Entrapment of microbial cells within polyurethane hydrogel beads with the advantage of low toxicity. Biotech. Techniques 6, 483-488.

Woias, P., Richter, M., Yacoub-George, E., Wolf, H., Abel, Th. (1996): A micromachined open tubular reactor for heterogeneous immunoassays. Analytical Methods & Instrumentation, Special Issure μTAS ´96.

6.2   Vitellogenin - A Biomarker for Endocrine Disruptors

P.-D. Hansen[1], H. Dizer[1], B. Hock[2], A. Marx[2], J. Sherry[3], M. McMaster[3] and Ch. Blaise[4]

[1] Berlin University of Technology, Dept. of Ecotoxicology, Keplerstraße 4-6, D-10589 Berlin, Germany

[2] B. Hock, A. Marx, Technical University of München, Department of Botany, Alte Akademie 12, D-85350 Freising, Germany

[3] J. Sherry, McMaster, National Water Research Institute, Burlington, Ontario, Canada

[4] Ch. Blaise, Centre Saint-Laurent, Montréal, Québec, Canada

6.2.1   Introduction

The increasing preponderance of females in fish populations has become a matter of concern not only for scientists but also for environmental authorities. For example, in the extensive waterways in Berlin, Germany, 70% of the fish population is female. A twofold question should be asked: is the biomarker vitellogenin (vitellogenin-synthesis assay) an appropriate tool for determining endocrine effects (European Commission 1996) and of chemical pollutants in effluents (Castillo et al. 1997), such as estradiols, phthalates, alkyl-phenols, and alkyl-ethoxylates, influence sex differentiation in fish?

The presence of vitellogenin in male fish was choosen as an indicator of exposure to extrogenic compounds. The vitellogenin protein is specific for each individual species and, thus, often requires the production of specific antibodies for the selected species (see Figure 71).

6.2.2 Materials and methods

Male fish exposed to effluents can, thus, be employed to monitor endocrine disruptions through multiple measurements of vitellogenin prodution as easily detected in their blood serum. The determination of vitellogenin is accomplished by means of a non-competitive enzyme immunoassay (EIA) using monoclonal antibodies.

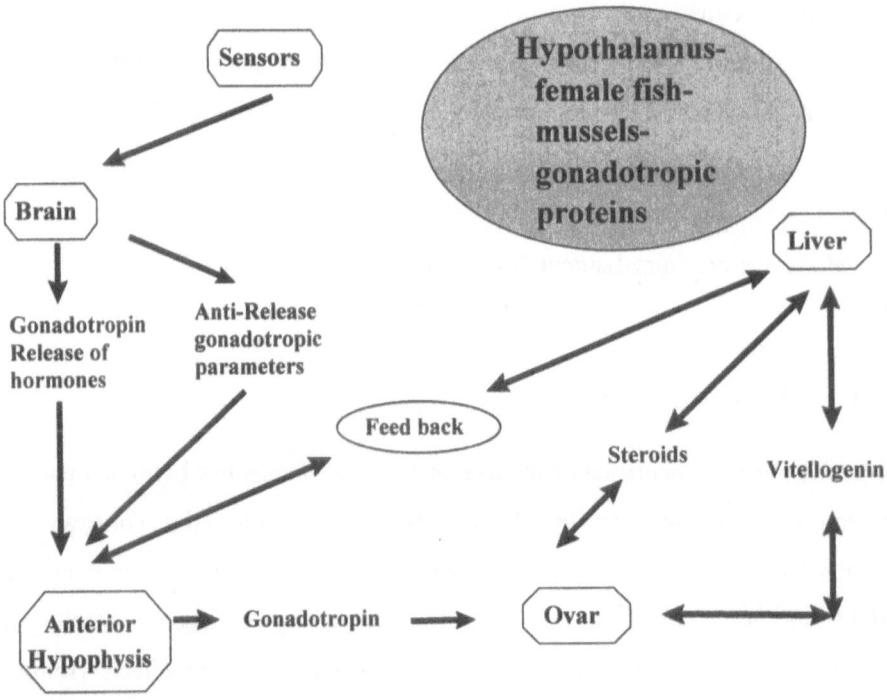

Figure 71: Vitellogenin synthesis in male fish as a summarizing effects parameter for the presence of endocrine disruptors

For the exposure experiments of fish in effluents the fish are exposed in the exposure tanks of a so called "on site WaBoLu-Aquatox Monitoring system" (Hansen 1986, 1988). The fish are exposed to a mixture of definite amounts of effluent and dilution water in the flow through system. The dilution steps (10%, 20% 30% and 40% effluent) are relevant concerning to the effluent loading of the Berlin waterways during the seasons of the year.

6.2.3   Results

The concentrations of vitellogenin produced in the effluent exposed male fish are shown in Figure 72. In additional experiments the endocrine effects of single contaminants of the effluents have to be investigated.

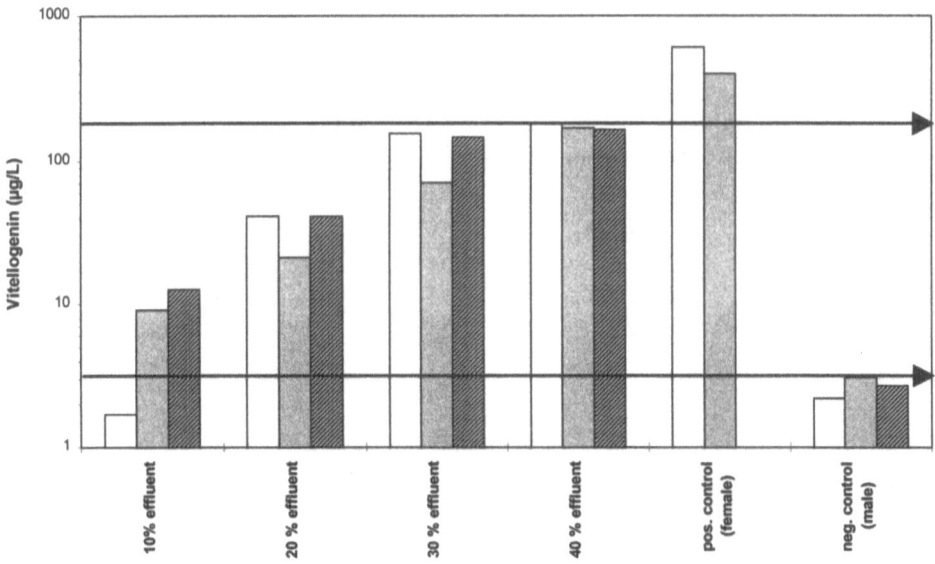

Figure 72:   Vitellogenin concentration in the serum of rainbow trouts exposed for 6 month to effluents (N = 30, each column represents the average of 10 trouts). The analyses were carried out by EIAs

In the Table 26a the kinetics of the induction of vitellogenin by 17 ß-estradiol is shown. The vitellogenin was induced by the injection of 17 ß-estradiol into the peritoneum of the male fish.

Table 26b shows the results of the production of vitellogenin in a primary fish cell line (hepatocytes, rain bow trout) by 17 ß-estradiol, nonylphenol, bisphenol A and Table 26c the data by a so called Indicator assay.

Table 26a: Induction of vitellogenin in male trouts by 17ß-Estradiol

Vitellogenin-Induction by 17 ß-estradiol 10 µg/l (N=10):		
5 days	3,9	mg/ml
10 days	26	mg/ml
15 days	93	mg/ml
21 days	114	mg/ml

Table 26b: Synthesis of vitellogenin by waste water contaminants with primary fish cells

Vitellogenin level 10 ng/ml (5 days):	
LOEC (LOEC=Lowest Observed Effective Concentration)	
17ß-Estradiol	**1 ng/l**
Nonylphenol	**14 µg/l**
Bisphenol A	**25 µg/l**

Table 26c: Endocrine effects determination by the MCF-7 ERE luciferase indicator assay

MCF-7 ERE Luciferase indicator-assay	
LOEC (LOEC=Lowest Observed Effective Concentration)	
Nonylphenol	**8 µg/l**
Bisphenol A	**12 µg/l**

Table 27: Endocrine contaminants (Gülden et al. 1997) in the effluents of the sewage plant Berlin-Ruhleben (29.08.96-19.09.97)

( ) = Limits of determination (LOD)

Hormones:			
Ethinyl-estradiol	>0,2-3,0	ng/l	(>0,2-<1,0 ng/l)
17 ß-Estradiol	>0,5-1,5	ng/l	(>0,1-<0,5 ng/l)
Estrone	>0,5 -<1	ng/l	(>0,1-<1,0 ng/l)
Phenols:			
4-tert-Octylphenol	> 30-110	ng/l	(30 ng/l)
4-Nonylphenol	> 80-923	ng/l	(80 ng/l)
Nonylphenolmonoethoxylate	> 50-138	ng/l	(50 ng/l)
Nonylphenoldiethoxylate	>110-213	ng/l	(110 ng/l)
Bisphenol A	8 - 33	ng/l	(2 ng/l)
Phthalates:			
Di-n-Butylphthalate	337-2,808	ng/l	(20 ng/l)
Benzyl-Butylphthalate	28-344	ng/l	(20 ng/l)
Di(-ethyl-hexyl)phthalate	622-40,007	ng/l	(20 ng/l)
Nitromusk:			
Musk-Xylene	20-32	ng/l	(20 ng/l)
Musk-Ketone	198-805	ng/l	(20 ng/l)
Pesticides:			
2,4'-DDT	<5	ng/l	(5 ng/l)
4,4'-DDT	<5	ng/l	(5 ng/l)
Lindane	19 - 28	ng/l	(2 ng/l)

The investigated contaminants of the effluents were selected from Table 27. Table 27 shows the results of the measurements of the contaminants in the effluents during the vitellogenin experiments and the exposed male fish.

6.2.4   Discussion

It is obvious that in effluent exposed male fish produce vitellogenin. Figure 72 shows clearly that there is a production of vitellogenin in male fish according to the effluent concentrations. There is already a remarkable increase of the vitellogenin in the serum of the fish exposed to > 20% effluent. In parallel the vitellogenin data of the none effluent exposed female fish (positive controls) and male fish (negative controls) were measured.

The kinetcs of the production of vitellogenin was demonstrated by induction experiments with 17 ß-estradiol. The vitellogenin was induced by an injection of 17 ß-estradiol into the peritoneum of the male fish. The results given in Table 26a shows that during an incubation time of 5 days there was already a remarkable increase of the vitellogenin in the serum of the male fish.

In cause effect studies with a primary fish celline and an indicator test the endocrine effects of selected single contaminants of the effluents were investigated. The vitellogenin concentration and the lowest observed effective concencentrations (LOEC) by the selected contaminants of the effluents are shown in Table 26b and 26c. For prediction studies the LOEC data have to be compared with the measured data of the contaminants in the effluents given in Table 27.

The nonylphenol concentrations in the effluents given in Table 27 are very low in comparison to the effects in the receptor- and indicator assays in the Tables 26b and 26c. There is a safety factor of more than 100 between the concentration in the effluents (Table 27) and the endocrine effect measured by the receptor and indicator assays (Table 26b and 26c). The safety factor concerning bisphenol A is approximately 3,000.

There is no safety factor (Hansen 1989) concerning the effects concentrations (LOEC) of the hormones and the measured concentrations of the hormones in the effluents. The endocrine effects of 17ß-estradiol shown in Table 26b were measured at a very low concentration (1 ng/l). These are the concentrations which are already measured in the effluents of the sewage plant.

In Table 28 the concentrations of the estrogenic substances in the effluents in Berlin-Ruhleben are compared with the concentrations of these estrogens in 17 sewage plants in the United Kingdom.

The concentations of the hormones in the effluents in Berlin are lower than those in the selected sewage plants in the United Kingdom. Overall the results show clearly that the LOEC (LOEC=Lowest Observed Effective Concentration) concerning the endocrine effects of 17ß-estradiol is very close to the environmental concentration in the effluents and consequently in the water ways.

The on-site exposure experiments with fish in the effluents of the sewage plant Berlin-Ruhleben and the cause effect studies with selected contaminants of the effluent gives rise to the further question of whether the increased vitellogenin synthesis (see Figure 72) in male fish is the correct answer to the sex ratio problem in the water ways.

Table 28: Estrogenes in sewage plant effluents in Berlin-Ruhleben and United Kingdom (N=17 sewage plants in UK)

Effluent Berlin - Ruhleben :			
Ethinyl-estradiol	>0,2–3,0	ng/l	(>0,2-<1,0 ng/l)
17ß-Estradiol	>0,5-1,5	ng/l	(>0,1-<0,5 ng/l)
Estrone	>0,5-<1	ng/l	(>0,1-<1,0 ng/l)
Effluents UK:			
Ethinyl-estradiol	0,2-7	ng/l	(>0,2-<1,0 ng/l)
17ß-Estradiol	4,0-48	ng/l	(>0,1-<0,5 ng/l)
Estrone	1,4-76	ng/l	(>0,1-<1,0 ng/l)

To answer this question, fish eggs from fish exposed to effluents were hatched and the fish larvae of the F2 generation were exposed in the on-site flow through system again. These experiments are in progress and the results may give more answers concerning the endocrine effects and the sex ratio of the fish stocks in the near future.

At this time the vitellogenin in the blood serum of the fish is measured by a Vitellogenin ELISA Assay. The directions for the future will be a stand alone biosensor enabling an on-site technique. As transducer and carrier a screen printed graphite electrode will be used.

6.2.5   Acknowledgement

The authors thank the Senatsverwaltung für Stadtentwicklung, Umweltschutz und Technologie (Wasserwesen) and the International Bureau of the BMBF under the Germany-Canada Science & Technology agreement for supporting the study. The development of the biosensor is supported under the Environment and Climate Programme of DG XII of the European Commission.

6.2.6   References

Castillo, M., Barceló, D. (1997): Analysis of endocrine-disrupting chemicals in industrial effluents. Trends in analytical chemistry TRAC, in press.

European Commission (1996): European Workshop on the Impact of Endocrine Disruptors on Human Health and Wildlife, 2-4 Dec.1996, Weybridge UK, Environment and Climate Research Programme, DG XII, EUR 17549, 125 pp.

Gülden, M., Turan, A., Seibert, H. (1997): Substanzen mit endokriner Wirkung in Oberflächengewässer, Umweltbundesamt Forschungsbericht 10204279, uba Texte 46, 361 pp.

Hansen, P.-D. (1986): Das "Wabolu-Aquatox" zur integralen Erfassung von Schadstoffen im Wasser.The "Wabolu-Aquatox" for Integral Monitoring of Water Pollutants. Vom Wasser 67, 221-235.

Hansen, P.-D. (1988): Wirkungsbezogene Biotestverfahren - Gefährliche Stoffe - Qualitätsziele zum Schutz oberirdischer Gewässer. Effect-related Bioassay Methods - Dangerous Substances - Quality objectives for the Protection of Surface Waters. Vom Wasser 70, 187-196.

Hansen, P.-D. (1989): Ecological requirements for the quality objectives in the aquatic environment, Tenside - Detergents 2, 80-84.

Hansen, P.-D., Schwanz-Pfitzner, I., Tillmanns, G.M. (1989): Ein Fischzellkulturtest als Ergänzungs- oder Ersatzmethode zum Fischtest. Bundesgesundheitsblatt 32, 8, 343-346.

7 Perspectives

7.1 How Could a Concerted Action Guide Technological Developments in the Field of Biosensors?

Alcock S.J. and Turner A.P.F.

Cranfield Biotechnology Centre, Cranfield, Bedfordshire, MK43 OAL, UK

Abstract. The need for guidance of technological developments in the field of biosensors for environmental monitoring is widely recognised. A Concerted Action entitled 'Biosensors for environmental monitoring' was recently funded under the EC Environment and Climate Programme. The aim of the Concerted Action is to enhance the development of biosensors for practical applications in monitoring environmental pollutants in air, water, soil and waste.

7.1.1   Introduction

The need for guidance of technological developments in the field of biosensors for environmental monitoring was recognised in 1996 at the 4th Workshop on Biosensors, a meeting of EC Environment and Climate grant holders and other interested parties. The participants elected a Science Panel which developed a proposal for an "umbrella project", a Concerted Action entitled 'Biosensors for environmental monitoring'. The aim is to enhance the development of biosensors for practical applications in monitoring environmental pollutants in air, water, soil and waste.

7.1.2   State of the art and industrial context

The quantity of chemicals released into the environment has risen dramatically in recent years, causing serious concern about their adverse effects on the ecosystem and on human health. The sources of pollution are numerous and the emissions are varied and often complex. Increasing environmental legislation which controls their release into the environment has created a need for reliable monitoring of these substances in water, air and soil. Many organisations in the public and private sectors are investing resources to develop fast and reliable monitoring methods (Bennetto and Büsing 1994). Biosensors offer unique advantages for environmental monitoring due to their specificity, fast response times, low cost, ease of use and continuous real time signal. The technology is capable of measuring both existing and new parameters of relevance to the environment, and of responding to a multiplicity of agents which are simultaneously present. In many monitoring situations biosensors can be expected to be the best available technology not entailing excessive costs.

In recent years biosensors have been successfully commercialised for clinical applications, but few have been marketed for environmental monitoring applications. There is a need for a closer link with industry, either with manufacturers or end-users (Barceló and Turner 1996). In order to address a large market, prospective biosensor devices will need to be either generic and detect the presence of a number of compounds or detect biological effects of pollutants, or alternatively capable of multianalyte detection and measurement. The devices need to be sufficiently robust to operate in a variety of environmental locations. There are a number of technical and commercial obstacles which must be addressed to allow biosensors to have a significant impact on environmental monitoring (Rogers 1995, Marco and Barceló 1996). The scientific and technical obstacles are the large number of potential pollutants, the broad range of chemical classes, the range and complexity of environment matrices and the consequent challenge of sample handling, the variety of possible chemically reactive co-contaminants and the wide dynamic range of pollutant concentrations. The commercial obstacles are lack of data requirement specifications, competition from other well-established and widely accepted methods, the need to identify sufficient large market sectors for specific applications, the necessity for rigorous validation studies on environmental samples and regulatory requirements. The purpose of the Concerted Action is to provide a forum to address these obstacles.

The advancement of biosensors for environmental monitoring requires scientists of different disciplines to work together to improve the transducer technology, to increase the number of analytes that can be measured, and to develop integrated units able to perform multi-analyte measurements (Marco and Barceló 1996). Strategies are required to enhance the specificity and sensitivity of biosensors for key individual analytes or classes of analyte for which current methods have limitations. The range of environmental monitoring situations where biosensors are useful can be extended e.g. to include operation on or in solid substrates, gases, industrial effluents and seriously perturbed ecosystems. Novel biosensors capable of measuring new parameters of relevance to the environment can be developed (Bennetto and Büsing 1994). For example biosensors could be developed to monitor endocrine-disrupting substances such as certain pesticides or hormones in industrial waste. All of these opportunities have recently been recognised in a series of national evaluations by Member States, resulting in the setting up of various Sensor Centres. This Concerted Action would draw together the local initiatives into a major European concerted effort to improve monitoring of our environment.

7.1.3 Objectives

The principal aims are to enhance the development of existing biosensors for routine application in environmental monitoring, to facilitate the creation of biosensors capable of measuring new parameters of relevance to the environment, to forward extension of the range of situations where biosensors may be applied to monitor the environment under extreme conditions, and to promote the development of integrated sensor systems capable of measuring several parameters simultaneously under real operational conditions. The main targets on which biosensor development will be focused are ground water, surface water, effluents, air, improvement of preventive technologies, control of waste disposal sites, and efficiency control of remediation activities. The Concerted Action involves four major tasks :

Task 1   : Involvement of industrial producers and users in addressing problems of mass production and of performance in real conditions, in order to bring to fruition the useful application of existing biosensors.

Task 2   : Enhancement of understanding of new analytical demands, including new parameters to describe environmental phenomena and effects, and the potential of biosensors to satisfy these demands.

Task 3   : Definition of specific environmental tasks as targets for biosensor development.

Task 4   : Identification of paths to resolve the most effective biosensors for use in monitoring integrated industrial waste (including air contaminants) and for use in waste water management in order to promote a clean environment.

7.1.4   Methodology

The Concerted Action will approach elimination of the gap between existing biosensor technology and its application to environmental monitoring by a number of complementary actions. The work will be based on a series of European meetings, a centralised information facility and a broad collaboration programme. This concertation will improve the amount of information available and promote information and technology exchange between academia, industry, end users and regulatory agencies, in the general area of biosensors for environmental monitoring. The Concerted Action will be guided by the elected Scientific Panel, and it will also involve all EC Environment and Climate grant holders, and other interested European partners, including specialist representatives from disciplines complementary to biosensor development (e.g. biochemists, molecular biologists, engineers and waste treatment technologists), industrial representatives and end users.

The tasks will be achieved by means of a series of specific objectives. The Concerted Action will focus European expertise in biosensors on the specific problems associated with environmental monitoring in real matrices. Tasks which are aimed in particular at improving the understanding of environmental processes and phenomena will be based on and accompanied by modelling of the respective processes. The matrices will be water, air, soil and waste. The programme will involve the development of interdisciplinary teamwork to identify the main research needs and the most appropriate approaches. It will include the identification of analytes in specific environmental matrices for which improved measurement techniques are required within the general categories of priority pollutants (EC "Black List" pollutants), volatile organic compounds (VOCs), microbial quality indicators, water quality parameters, gases and phenols. The Concerted Action will promote the characterisation of the sensor surface in contact with the environmental matrix both prior to and after exposure. Novel materials (e.g. antibodies) and prototype devices will be exchanged for evaluation and comparison.

The Concerted Action will identify supporting R&D tools of European interest, e.g. antibody libraries, field measurement campaigns. It will review and exchange techniques for the construction of sufficient numbers of sensors for statistical evaluation for signal processing and the production of appropriate electronic configurations.

A forum will be provided for exchange of information, involving universities, research centres, manufacturers, and end users in order to facilitate the transfer of knowledge to industry and others and to identify gaps in knowledge and the respective R&D projects to address them. This will be underpinned by facilitating the performance of market studies for biosensors for environment monitoring in areas where industrial support is evident. Required sensor performance will be defined with regard to environmental limitations. The Concerted Action will document the advantages over alternative assays of monitoring using biosensors. It will harmonise methods of calibration and reference methods, organise testing facilities and define standard procedures for evaluation of results with particular regard to effects of the environmental matrix. The requirements for biosensors for environmental monitoring will be presented to interested academic and commercial organisations in Europe.

7.1.4.1   European meetings

Three major workshops will be held involving all EC Climate and Environment: Biosensors for Environmental Monitoring grant holders, participating workers and other European researchers from various relevant disciplines and guest speakers. Members will be asked to report on relevant national, EU and, where possible, industrial projects. The function of these meetings will be to share information, to discuss critical issues and to formulate environmental testing protocols. Where appropriate these meetings will include demonstration sessions and involve contributions from industry. The main workshops will address the following areas :

Technology Evaluation: Identification of environmental monitoring problems. Definition of monitoring tasks. Limitations in application of existing biosensors, receptors and transducers to monitor pollutants in specific environmental matrices. Identification of technologies which are most promising for environmental monitoring. Strategies to enhance performance of biosensors. Development of integrated sensor systems for simultaneous measurement of multiple parameters.

Response to New Analytical Demands: New parameters to describe environmental phenomena and effects. Definition of new monitoring tasks. Potential of biosensors to meet new demands. Specific technologies which can be applied. Characterisation of the sensor surface. Calibration and reference methods. Simultaneous measurement of multiple analytes. Modelling environmental and remediation processes.

Monitoring in Real Environments: Technological advances. Practical achievements. Effects of the environmental matrix. Sensor performance under extreme conditions.

In addition to the above, four smaller workshops will focus on technologies for monitoring respectively in the environmental matrices water, air, soil and waste. They will all report into the final workshop on 'Monitoring in Real Environments'.

7.1.4.2 Access to the Concerted Action

A home page will be set up on the World Wide Web to inform the scientific community of the new strategy. The web site will include private pages available to registered academic and industrial European users. These will contain more detailed information e.g. data gathered via questionnaires such as details about individual research groups, availability of antibodies, facilities and opportunities for exchange of personnel, environmental monitoring tasks and details of sensors under development in Europe. A regular edited newsletter will be published via the Internet, containing information relating to both the participants in the Concerted Action and other researchers in the field. The publication of experimental data and the result of workshops will be actively pursued by producing workshop proceedings and articles for scientific journals. At the conclusion of the Concerted Action a glossy brochure summarising the state of the art will be produced for wide distribution.

Short visits will be sponsored for researchers from European laboratories active in the field to learn techniques, use equipment and consult with other groups. Access to specialist equipment and expertise will be organised, including testing in common or standardised environmental samples, statistical advice, microfabrication and encapsulation techniques, computer facilities, electronics and instrument design, synthetic chemistry and enzyme technology. Exchange of workers and technology transfer between academia and industry will be actively promoted. Individual and collaborating groups will be encouraged to patent their inventions and to seek joint exploitation whenever it is beneficial.

Assistance will be provided to establish joint projects between European research groups and European industry. They will be designed to solve well-defined problems related to biosensors for environmental monitoring. The Concerted Action will lend support in the formulation of research proposals and in seeking funding mechanisms. This will include a proposal under the EC Environment and Climate Accompanying Measures programme for a small scale field experiment to evaluate and compare biosensors in a real environment.

The Project Steering Group will inform and collaborate with related complementary EU Actions e.g. the European Environment Agency (EEA), NICOLE Concerted Action (Network for Industrially Contaminated Land in Europe) CARACAS (Concerted Action on Risk Assessment for Contaminated Sites in the EU), LCANET (European Network for Strategic Life Cycle Assessment), Brite-Euram III Concerted Action on Biosensor Stability, the proposed Concerted Action on Environmental Technology (ETCA) and other international activities e.g. the European Science Foundation. The Group will assist bodies such as the European Committee for Standardisation (CEN) and IUPAC (Electrochemical Biosensor Working Party and ISE Group) in the formulation of standards, guidelines and nomenclature.

The Project Steering Group will be charged with the responsibility of maintaining close contact with European industry and actively exploring possibilities of collaboration with the commercial sector. This will ensure that the eventual transfer of the technology to marketable devices is always a major consideration in the programme and that such transfer is achieved as smoothly and effectively as possible.

7.1.5   Conclusion

Future advances in biosensor development will require scientists of different disciplines to join their research efforts to reach important goals. The Concerted Action will be a bridge of exchange and communication. Co-ordination of the highly skilled, but dispersed European activity in this field will undoubtedly enhance performance of existing biosensors, and lead to the development of new and better devices. It will provide the opportunity to pool resources, perform comparative experiments, rapidly disseminate information and lead to the development of new national initiatives in the area. There will be tangible benefits from bringing individual programmes together, avoiding the duplication of isolation and facilitating faster solution of technical problems. Thus it is expected that this will add considerable value to existing projects, and that a number of biosensor systems will be evaluated for monitoring in environmental matrices and that the results will be made available.

The Concerted Action will elucidate new needs and the most effective analytical solutions to problems in waste disposal, control of clean technology and water resources. It will facilitate the development of practical, selective, low cost, rapid, durable, multi-analyte biosensor devices.

7.1.6   Acknowledgements

This work is based on the contributions of the Science Panel:

Dr. D. Barceló, CID-CSIC, Barcelona, Spain; Dr. J. Büsing, Scientific Officer, European Commission, Professor E. Dominguez, University of Alcala de Henares, Madrid, Spain; Professor L. Fiksdal, Norwegian University of Science and Technology, Trondheim, Norway; Professor P.-D. Hansen, Technische Universitat Berlin, Germany; Dr. D. Nowak, Nowak Umweltanalysen, Berlin, Germany; Ms. P. Sztajnbok, Anjou Recherche, Maisons-Laffitte, France; Professor A.P.F. Turner, Cranfield University, Bedford, UK.

7.1.7   References

Barceló D., Turner A.P.F. (1996): Fourth European Workshop: Biosensors for Environmental monitoring. EC Report. ISSN 1018-5593. 193pp.

Bennetto P., Büsing J. (eds.) (1994): Second European Workshop: Biosensors for Environmental monitoring. EC Report EUR 15622EN. ISSN 1018-5593. 245pp.

Marco, M.P., Barceló D. (1996): Environmental applications of analytical biosensors. Meas. Sci. and Technol. 7, 1547-1562.

Rogers, K.R. (1995): Biosensors for Environmental applications. Biosensors and Bioelectronics 10, 533-541.

Subject index

274

Authors

Abuknesha, Dr. R. A.
Division of Life Science
King's College
Kensington
Campden Hill Road
London W8 7AH
UK
Phone: +44 171 333 4483
Fax: +44 171 333 4500
E-mail: ram.abuknesha@kcl.ac.uk

Barceló, Prof. Damià
Departament de Química Ambiental
C.I.D.-C.S.I.C.
Jordi Girona, 18-26
E-08034 Barcelona
Spain
Tel.: +34 3 4006118
Fax: + 34 3 2045904
E-mail: dbcqam@cid.csic.es

Barzen, Claudia
Institute of Physical Chemistry
University of Tübingen
Auf der Morgenstelle 8
D-72076 Tübingen
Germany
Phone: +49 7071 2974667
Fax: +49 7071 296910
E-mail: claudia.barzen@ipc.uni-
tuebingen.de

Brecht, Dr. Andreas
Institute of Physical Chemistry
University of Tübingen
Auf der Morgenstelle 8
D-72076 Tübingen
Germany
Phone: +49 7071 2978753

Fax: +49 7071 296910
E-mail: andreas.brecht@ipc.uni-
tuebingen.de

Büsing, Dr. Jürgen
European Commission
DG XII-D-1
SDME 7/56
Rue de la Loi 200
B-1049 Brussels
Belgium
Phone: +32 229 55625
Fax: +32 229 63024

Cammann, Prof. Dr. Karl
Institut für Chemo- und Biosensorik
e.V., ICB
Mendelstr. 7
D-48149 Münster
Germany
Phone: +49 251 980 2800
Fax: +49 251 980 2802
E-mail: Cammann@uni-muenster.de

Castillo, Dr. M.
Departament de Química Ambiental
C.I.D.-C.S.I.C.
Jordi Girona, 18-26
E-08034 Barcelona
Spain

Del Carlo, Michele
Dpt. Sanità Publica
Sez. Chimica Analitica
Via Gino Capponi 9
50121 Firenze
Italy
Phone: +39 55 2757265
Fax: +39 55 2476972

E-mail: mascini@cesit1.unifi.it

Dominguez, Prof. Dr. Elena
Department of Analytical Chemistry
Faculty of Pharmacy,
University of Alcalá,
E-28871 Alcalá de Henares (Madrid)
Spain
Phone +34 1 885 4666;
Fax +34 1 885 4666/0
E-mail qadominguez@ixnqui.alcala.es

Fraval, Soizic
Anjou Recherche-Le Graal
Rue de Cotes 109-111
78600 Maisons-Laffitte
France
Phone: +33 1 3493 3455
Fax: +33 1 3493 3469

Gascon, Dr. J.
Departament de Química Ambiental
C.I.D.-C.S.I.C.
Jordi Girona, 18-26
E-08034 Barcelona
Spain

Gauglitz, Prof. Dr. Günter
Institute of Physical Chemistry,
University of Tübingen
Auf der Morgenstelle 8
D-72076 Tübingen
Germany
Phone: +49 7071 2976927
Fax: +49 7071 296910
E-mail: guenter.gauglitz@uni-
tuebingen.de

Hansen, Prof. Dr. Peter-Diedrich
TU Berlin, FB7
Institute of Ecology
Dept. of Ecotoxicology

Keplerstr. 4-6
D-10589 Berlin
Germany
Phone: +49 30 31421463
Fax: +49 30 8318113
E-mail: pd.hansen@tu-berlin.de

Harris, Dr. Richard
Optoelectronics Research Centre
University of Southampton
Southampton,
Hants 8017 IBJ
UK
Tel: +44 1703 593162
Fax: +44 1703 593149
E-mail: rdh@orc.soton.ac.UK

Hock, Prof. Dr. Bertold
Dept. of Botany
TU München at Weihenstephan
D-85350 Freising
Germany
Phone: +49 8161 713395
Fax: +49 8161 714403
E-mail: hock@pollux.edv.agrar.tu-
muenchen.de

Horneck, Dr. Gerda
DLR Institut für Luft- und
Raumfahrtmedizin
D-51170 Köln
Germany
Phone: +49 2203 601 3594
Fax: +49 2203 61970
E-mail: gerda.horneck@dlr.de

Katakis, Dr. Ioanis
University Rovira i Virgili de
Tarragona
Escola Tecnica Superior d'Énginyeria
Quimica
E-43006 Tarragona Catalonia

Spain
Phone +34 7755 9655;
Fax +34 7755 9621
E-mail: ikatakis@etseq.urv.es

Kramer, Dr. Karl
TU München at Weihenstephan
D-85350 Freising
Germany
Phone: +49 8161 713874
Fax: +49 8161 714403
E-mail: kramer@pollux.edv.agrar.tu-muenchen.de

Marco, Dr. Maria-Pilar
Department of Biological Organic
Chemistry
C.I.D.-C.S.I.C.
Jordi Girona, 18-26
E-08034 Barcelona
Spain

Mascini, Prof. Marco
Dpt. Sanità Publica, Sez. Chimica
Analitica
Via Gino Capponi 9
50121 Firenze
Italy
Phone: +39 55 2757265
Fax: +39 55 2476972
E-mail: mascini@cesit1.unifi.it

Meusel, Dr. Markus
Institut für Chemo- und Biosensorik
e.V., ICB
Mendelstr. 7
D-48149 Münster
Germany
Phone: +49 251 980 2879
Fax: +49 251 980 2890
E-mail: Meusel@uni-muenster.de

Morey, Dr. Johanna
Bookham Technology Ltd.
Rutherford Appleton Laboratory
Chilton
Oxfordshire OX11 OQX
UK
Phone: +44 1235 445377
Fax: +44 1235 446854
E-mail: johannam@bookham.co.uk

Oubiña, Dr. A.
Departament de Química Ambiental
C.I.D.-C.S.I.C.
Jordi Girona, 18-26
E-08034 Barcelona
Spain

Santandreu, Marta
Universitat Autònoma de Barcelona
Grup de Sensors i Biosensors
08193 Bellaterra, Barcelona
Spain
Phone: +343 581 21 18
Fax: +343 581 24 77

Schmid, Prof. Dr. Rolf
Institute for Technical Biochemistry
University of Stuttgart
Allmandring 31
D-70569 Stuttgart
Germany
Phone: +49 711 685 3193
Fax: +49 711 685 3196
E-mail: r.schmid@po.uni-stuttgart.de

Seifert, Martin
Dept. of Botany
TU München at Weihenstephan
D-85350 Freising
Germany
Phone: +49 8161 713973
Fax: +49 8161 714403

E-mail: seifert@pollux.edv.agrar.tu-
muenchen.de

Stöcklein, Dr. Walter
Universität Potsdam
Institut für Biochemie und Mole-
kulare Physiologie
c/o Max-Delbrück-Centrum
(MDC)
Robert-Roessle-Str. 10
D-13122 Berlin
Germany
Phone: +49 30-9489-2650
Fax: +49 30-9489-3322
E-mail: wstoeck@mdc-berlin.de

Sztajnbok, Pascale
Anjou Recherche

Chemin de la Digue
BP 76
78603 Maisons-Laffitte
France
Phone: +33 1 3493 3110
Fax: +33 1 3493 3111

Turner, Prof. Anthony P. F.
Cranfield Biotechnology Centre
Institute of BioScience and
Technology
Cranfield University
Cranfield
Bedfordshire MK43 OAL
UK
Phone: +44 1234 754132
Fax: +44 1234 752401
E-mail: a.p.turner@cranfield.ac.uk

Stoyan/Stoyan/ Jansen
Umweltstatistik

Statistische Verarbeitung und Analyse von Umweltdaten

Von Prof. Dr.
Dietrich Stoyan
Helga Stoyan
und Dr.
Uwe Jansen
Technische Universität
Bergakademie Freiberg

Umweltforschung ist ohne Statistik nicht denkbar. Viele Beobachtungen, Messungen und Versuche führen zu riesigen Datenmengen, die ohne statistische Auswertung nutzlos wären. Weil diese Daten inhomogen, zeitabhängig, räumlich und hochdimensional sind, werden Analyseverfahren benutzt, die in Anfängervorlesungen nicht geboten werden können: Multivariate Statistik, Zeitreihenanalyse, Geostatistik, Punktprozeß-Statistik und Extremwertstatistik. Diese Verfahren werden hier praxisnah anhand von Beispielen erläutert. Dabei geht es um geochemische, hydrogeologische und meteorologische Fragestellungen, um Luftverschmutzung, Abfallwirtschaft und Altlastenuntersuchungen.

1997. 348 Seiten.
16,2 x 22,9 cm.
Kart. DM 64,80
ÖS 473,– / SFr 58,–
ISBN 3-8154-3526-9

(Teubner-Reihe UMWELT)

Preisänderungen vorbehalten.

B. G. Teubner Stuttgart · Leipzig